# COMMERCE 3M

## TABLE OF CONTENTS & ACKNOWLEDGEMENTS

| | PAGE |
|---|---|
| Marketing Strategy and Tactics<br>    Ryder, M.<br>    © 2016 Ryder, Prof. Marvin G.<br>    Reprinted with permission. | 1 |
| Market Segmentation<br>    Ryder, M.<br>    © 2016 Ryder, Prof. Marvin G.<br>    Reprinted with permission. | 17 |
| Product Strategy<br>    Ryder, M.<br>    © 2016 Ryder, Prof. Marvin G.<br>    Reprinted with permission. | 25 |
| Pricing Strategy<br>    Ryder, M.<br>    © 2016 Ryder, Prof. Marvin G.<br>    Reprinted with permission. | 37 |
| Promotion Strategy<br>    Ryder, M.<br>    © 2016 Ryder, Prof. Marvin G.<br>    Reprinted with permission. | 45 |
| Distribution Strategy<br>    Ryder, M.<br>    © 2016 Ryder, Prof. Marvin G.<br>    Reprinted with permission. | 57 |
| Introduction to the Case Method<br>    Ryder, M.<br>    © 2016 Ryder, Prof. Marvin G.<br>    Reprinted with permission. | 69 |
| Marketing Research<br>    Ryder, M.<br>    © 2016 Ryder, Prof. Marvin G.<br>    Reprinted with permission. | 81 |

Consumer Behaviour     91
    Ryder, M.
    © 2016 Ryder, Prof. Marvin G.
    Reprinted with permission.

International Marketing     103
    Ryder, M.
    © 2016 Ryder, Prof. Marvin G.
    Reprinted with permission.

Crayola Canada Ltd.     111
    Ryder, M.
    © 2016 Ryder, Prof. Marvin G.
    Reprinted with permission.

Lime Light Cinema     119
    Ryder, M.
    © 2014 Ryder, Prof. Marvin G.
    Reprinted with permission.

Thompson Brothers Funeral Homes     125
    Ryder, M.
    © 2015 Ryder, Prof. Marvin G.
    Reprinted with permission.

East Hamilton Miniature Golf     131
    Ryder, M.
    <u>East Hamilton Miniature Golf</u>, Ryder, M.
    © 2006 Ryder, Prof. Marvin G.
    Reprinted with permission.

Julius Schmid of Canada Limited     139
    Ryder, M.
    <u>Marketing Insights: Contemporary Canadian Cases</u>, Ryder, M.
    © 2003 Ryder, Prof. Marvin G.
    Reprinted with permission.

Huron Canvas Clothier     147
    Ryder, M.
    © 2015 Ryder, Prof. Marvin G.
    Reprinted with permission.

Forum des Arts     153
    Ryder, M.
    <u>Forum des Arts</u>, Ryder, M.
    © 01/01/1994 Ryder, Prof. Marvin G.
    Reprinted with permission.

Dominion Tanking Limited 161
    Ryder, M.
    Ryder, M.
    © 2014 Ryder, Prof. Marvin G.
    Reprinted with permission.

E.D. Smith and Sons Limited 165
    Ryder, M.
    Marketing Insights: Contemporary Canadian Cases, Ryder, M.
    © 2000 Ryder, Prof. Marvin G.
    Reprinted with permission.

Tremco Ltd. 173
    Ryder, M.
    Tremco Ltd., Ryder, M.
    © 2017 Ryder, Prof. Marvin G.
    Reprinted with permission.

National Music Studio 181
    Ryder, M.
    National Music Studio, Ryder, M.
    © 2017 Ryder, Prof. Marvin G.
    Reprinted with permission.

Porsche Care Canada Ltd. 189
    Ryder, M.
    Porsche Care Canada Ltd., Ryder, M.
    © 2017 Ryder, Prof. Marvin G.
    Reprinted with permission.

Fortron International Inc. 197
    Ryder, M.
    Fortron International Inc., Ryder, M.
    © 2017 Ryder, Prof. Marvin G.
    Reprinted with permission.

# MARKETING - STRATEGY AND TACTICS

Every textbook defines marketing differently. For our purposes, marketing is the art of finding out what people want and giving it to them in a way better than the competition, while generating revenue over the long term. There are, of course, some limitations to this definition. "Revenue" connotes "dollars and cents" yet non-profit organizations do engage in marketing. Revenue should, therefore, be more broadly considered as a response – it may be the number of pints of blood donated, attendance figures at a little-league game, or even the amount of money a foundation donates to others. The timing of revenues is also important. Successful marketers look for revenue over the long term. It may be necessary, for a year or two, to have lower revenues to fully satisfy consumers. When Johnson & Johnson was faced with the discovery of poisoned capsules in some bottles of its best-selling pain reliever, Tylenol, it ordered a worldwide recall of the product costing hundreds of millions of dollars. While it did not generate revenue for the company in the short term, this action helped ensure revenues for Johnson & Johnson over the long term by inspiring trust in the consumer.

Another limit to the definition is that some things that people want they should not have. There are members of society who want access to prostitution, drugs and child pornography. The rest of society imposes stiff penalties on the promotion and selling of these commodities. The inverse is also true: some things people do not want they should have. Many people don't want to wear seat belts in automobiles or safety helmets while riding motorcycles, but the rest of society has demanded that these precautions be promoted. Only serving consumer wants can lead to tunnel vision. Consumers are not well versed in technological capabilities – they cannot see large technological leaps. Relying strictly on consumers, companies might not have envisioned compact-disc players, personal computers, or the Internet.

Though elements of marketing can be traced to ancient Rome, the first marketing course did not appear until the early 1900's. Why? For marketing to exist, consumers must face a surplus of goods. In Somalia, Ethiopia, and parts of the former Soviet Union, there is a shortage of goods. We call that a seller's market. Finding out what consumers want is a mostly pointless exercise if a constant flow of product cannot be guaranteed. So it was in North America until the end of the Second World War. The assembly line and corresponding mass-production technology were not invented until the twentieth century, and were not perfected until the 1940's. As the soldiers returned home, there was, for the first time, a surplus of goods, or a buyer's market. When we move from a seller's market to a buyer's market, power is transferred from the producer to the consumer. A company that pays close attention to the needs of consumers can obtain a competitive advantage.

Ask the general population about marketing and the words you are most likely to hear in response are "sales" and "advertising." For many people, selling is a negative activity, which implies that consumers are forced or tricked into buying something that they do not want. Others view advertising more positively, but place a remote-control unit in the hands of someone watching a television program and you will soon see how much people dislike advertising. Marketing is, therefore, not viewed very favourably by the public. Perhaps if the marketing concept was understood in a broader context, marketing would gain a better reputation.

## ◊◊ The Four P's of Marketing

In battle, one force fights for territory using the weapons of its army, navy, air force and marines. In business, one company fights for consumer purchases using the four P's of marketing: product, promotion, price and place (or distribution) (see fig. 1). The concept of product goes far beyond the physical product itself, and includes package design, brand names, trademarks, warranties, guarantees, product life cycles, and new-product development. The product concept also includes the increasingly important service sector. Pricing begins with a consideration of economic principles

*Figure 1.* THE FOUR P'S OF MARKETING

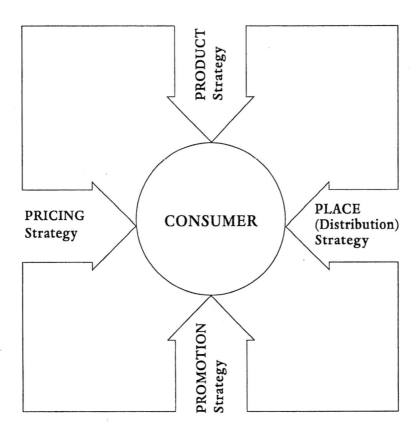

SOURCE: Dale Beckman, David Kurtz, and Louis Boone, *Foundations of Marketing*, 4th ed. (Toronto: Holt, Rinehart and Winston of Canada, 1988), 14.

*MARKETING INSIGHTS*

and then takes into account consumer expectations, manufacturing costs, mark-ups, discounts, and transportation costs. Promotion represents all forms of communication to consumers including advertising, social media, personal selling, public relations, and sales- promotion techniques (that is, coupons, samples, point-of-purchase displays). Place includes the activities of wholesalers and retailers, as well as choices involving the physical distribution of the product, and storage and inventory handling.

◊◊ The Environment for Marketing Decisions

Even with good weapons, superior forces have been beaten in battle by inferior forces. Why? No doubt the superior forces did not do a full reconnaissance and failed to consider the uncontrollable forces during battle. Many battles have been won or lost because of the weather or the unforeseen existence of a swamp. Marketers must realize that their deci-

*Figure 2.* THE ENVIRONMENT FOR MARKETING DECISIONS

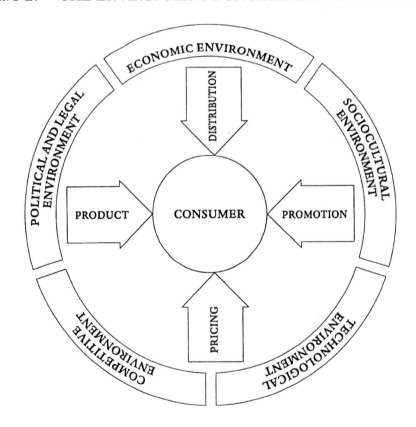

SOURCE: Dale Beckman, David Kurtz, and Louis Boone, *Foundations of Marketing*, 4th ed. (Toronto: Holt, Rinehart and Winston of Canada, 1988), 22.

sions involving the four P's are made in a context where the environment is uncontrollable. While marketers may influence some of these environmental factors, control eludes them (see fig. 2).

The competitive environment has become increasingly important as trade becomes more globalized. A company must be concerned with other firms in the industry, other firms that compete for the same customers, and it must contend with potential new competitors, both national and international. McDonald's Restaurants closely watches Burger King, Wendy's, and Harvey's, its competition in the hamburger industry. It also watches Pizza Hut, Tim Horton's, and Swiss Chalet, which compete for the same quick-service customers. McDonald's is also watching those quick-service companies from other parts of the world that would like to enter the affluent North American market.

The economy operates in a cyclical manner. Over a ten-year period, a marketer will see periods of recession, depression, recovery, and prosperity. Unemployment, inflation, and the economy in general influence many consumer decisions, including those concerning marriage, having children, buying a home, buying a car, and retirement. Closely connected is the political/legal environment. Government rules and regulations at the municipal, provincial, and federal level affect pricing (for example, the harmonized sales tax – HST), promotion (advertising to children, misleading advertising), distribution (where and how liquor, wine, and beer can be sold) and product (listing of ingredients, defining product terms). Society expects governments to control business and maintain a free-enterprise system. The rules of the Competition Act come into play here as they seek to prevent actions which greatly lessen competition (including mergers and acquisitions) and to prevent deceptive trade practices. Of course, a nation's laws are not static and can change as governments are replaced. As well, with increased emphasis on international trade, the political/legal environments of other nations become much more important.

Technology refers to knowledge gained from scientific investigation, discovery, and invention. It affects marketing in many ways. Technological advancement is the source of new products and product refinements. Through improved production efficiency, costs can be lowered and the savings shared with consumers. Technological advances in animation and sound recording have changed the production of television commercials. The universal product code and scanner technology have improved inventory handling and reduced the time consumers spend at retail checkouts.

The final environment looks at the interpersonal relationships of humans through their societies and cultures. As cultures and sub-cultures evolve, demand for products change and new products are needed. As Canada has welcomed immigrants from around the world, the food options at restaurants have evolved from meat with potatoes to quesadillas, pad Thai, and kimchee. More importantly, as we engage in global trade, the cultures of different countries need to be explored and respected. Similarly, understanding how consumers behave and how they choose products to purchase can improve the chances of success for a marketer.

## ◊◊ Developing a Marketing Strategy

A CFL football player and his wife asked for my opinion on a new business venture. He was an offensive lineman: three hundred pounds of muscle standing six feet six inches with a twenty-inch neck, looking for all the world like a brick wall. He liked fashionable clothing, but had a problem finding things that would fit. "So what do you think about us opening a fashionable big and tall man's clothing store?" he asked. "I would be personally involved in the off-season, and it would give me some security for when I retire in a few years."

"I think it could work," I replied. "But I need a few details."

"We see the store operating in about six hundred square feet in a major downtown shopping mall. We would carry some casual pants and shirts, belts, some sports jackets and suits. And great labels like Polo, Gant, Hugo Boss, and Armani."

"Sounds good to me," I said, "and I assume you will be using your celebrity status to draw customers."

"Don't forget his teammates," his wife added. She, too, was striking. She stood five feet eleven inches and had dabbled in bodybuilding. "I thought that we would probably have enough room in the store to include some clothes for the large and tall woman."

"Oh?" I was starting to get worried.

"Yes, and don't forget the shoes. I have a size fourteen foot, and I can't find good-quality shoes in my size," the football player chimed in.

"Men's and *women's* shoes," she added. "And, of course, we would need some accessories. Handbags, hats, some jewellery. I was even thinking of watches."

"Of course," was all I could say.

"And we would need to have coats for the fall and winter. Maybe some ties, and then there would be men's jewellery," he said.

I was flabbergasted. In thirty seconds, this couple had described a big-and-tall, high-class department store crammed into six hundred square feet. I knew this strategy wouldn't work. They were trying to satisfy too many consumers and too many needs at once. As the saying goes, a jack-of-all-trades is a master of none. Translation? A company that tries to do too many things at one time is likely to do none of them well.

As far as I could tell, I only had one problem. How would I tell a three-hundred-pound CFL lineman that his strategy was doomed to fail?

## ◊◊ Strategy and Tactics

*Strategy* is difficult to define. In an academic context, it can be described as the plan used to determine the direction of an organization and to achieve its long-term goals. When a company enacts its marketing strategy, it blends product, pricing, distribution, and promotion decisions to satisfy its chosen target market over the long term. As its reward for satisfying its customers, the company receives sales revenue and, ideally, sees profits. Strategic plans are often quite broad, they use a three-to-five-year time frame, and they encompass major overall objectives. Strategy has always been important in a military context. For instance, in the Gulf War of the early 1990's, the long-term military goal of the United Nations forces was the liberation of Kuwait. On the market battleground, the goal of a company might be to introduce its product to ten new countries or to create awareness of a new product in 50-percent of the population in a three-year period.

*Tactics* are narrower in scope than strategies. They focus on the implementation of those activities specified in the strategic plan. In fact, two different companies might use completely different tactics to accomplish the same strategic goals. Suppose the strategic goal is to introduce a product to ten new countries in three years. One company might choose to employ an export strategy that includes manufacturing the product in Canada and signing agreements with distributors in the new countries. Another company might purchase a factory in each of the ten countries to save on shipping costs and to exploit cheaper local raw materials. Generally, tactics are reviewed every six months, focus on short-term accomplishments, and involve actual resource allocation.

Everyone in an organization does some strategic and tactical planning, though the percentage of time an employee spends on these activities increases as that person moves to higher echelons of management. There is a long-standing debate about whether strategy formulation is an art or science. Those who think it is a science have spent considerable time developing rigorous techniques with which to turn specified inputs into generic strategies. Let's examine a couple of these techniques or strategy models.

## ◊◊ Generic Approaches to Strategy

One of the first generic approaches to strategy was championed by General Electric (GE) in 1971. With sales revenue measured in the billions and plants situated around the world, GE needed to focus its strategic planning. The first step in this process was to form strategic business units – groupings of companies that serve the same consumers with products requiring much the same resources to produce. The most important innovation from a strategic-planning perspective was the development of the GE business screen (see fig. 3).

There are two dimensions to this screen. One dimension is an internal measure called business strengths, and includes an assessment of the strength of the strategic business unit in the areas of production, human resources, information systems, finance, marketing, and research and development. If a unit shows outstanding strengths, it is rated high on the business-strengths dimension. The other dimension is an external measure called industry attractiveness, and includes an assessment of the competition, the sales growth rate, consumer behaviour, technological developments, government intervention, and economic conditions. If the environment facing a strategic business unit is quite threatening, industry attractiveness is rated as low.

Once these dimensions are plotted on the screen, one of three generic strategies is suggested. Where both dimensions received a high rating, the company should invest money in making the business unit grow. The future-earnings potential of the company lies in such a unit. Where both dimensions received a medium rating, the strategy should be more cautious. Management is told to be more selective about investing money in these units, and to do so only if the investment is made to support the current-earnings potential of the unit. Where both dimensions received a low rating, management is advised to cease investing and to work at squeezing every last ounce of profits from the unit. If the unit is not profitable or cannot be made profitable in short order, management is told to sell off or close the unit.

*Figure 3.* THE GE BUSINESS SCREEN

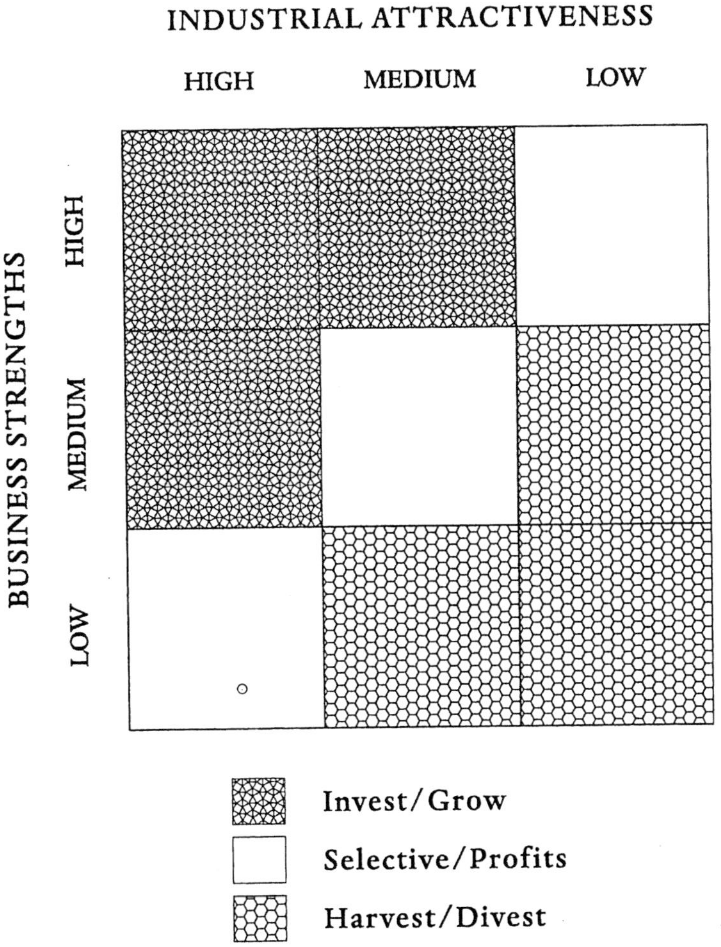

INDUSTRIAL ATTRACTIVENESS

|  | HIGH | MEDIUM | LOW |
|---|---|---|---|
| **HIGH** | Invest/Grow | Invest/Grow | Harvest/Divest |
| **MEDIUM** | Invest/Grow | Selective/Profits | Harvest/Divest |
| **LOW** | Selective/Profits | Harvest/Divest | Harvest/Divest |

BUSINESS STRENGTHS

- Invest/Grow
- Selective/Profits
- Harvest/Divest

SOURCE: Dale Beckman, David Kurtz, and Louis Boone, *Foundations of Marketing*, 2d ed. (Toronto: Holt, Rinehart and Winston of Canada, 1982), 131.

One problem with this screen is that it is supposed to be applied to strategic business units only. Not every company is large enough to be organized around to the business-unit concept. However, companies have successfully used the screen for individual products or target markets. Another set of problems involved the differences between high and medium, and between medium and low. Companies found that managers can interpret them quite differently

In the late 1970's, it was determined that a simpler screen was needed. A group of Harvard business professors, working through their independent consulting company, proposed a new model. Known today as the Boston Consulting Group (BCG) Matrix, it, too, has both an external and internal dimension (see fig. 4).

The external dimension is market growth. The assumption is that for high market growth to occur, technology, competition, economic conditions, the government, the laws, and the customers must all be favourable. As well, only one decision need be made, and that is whether market growth is high or low.

*Figure 4.*   THE BOSTON CONSULTING GROUP MATRIX

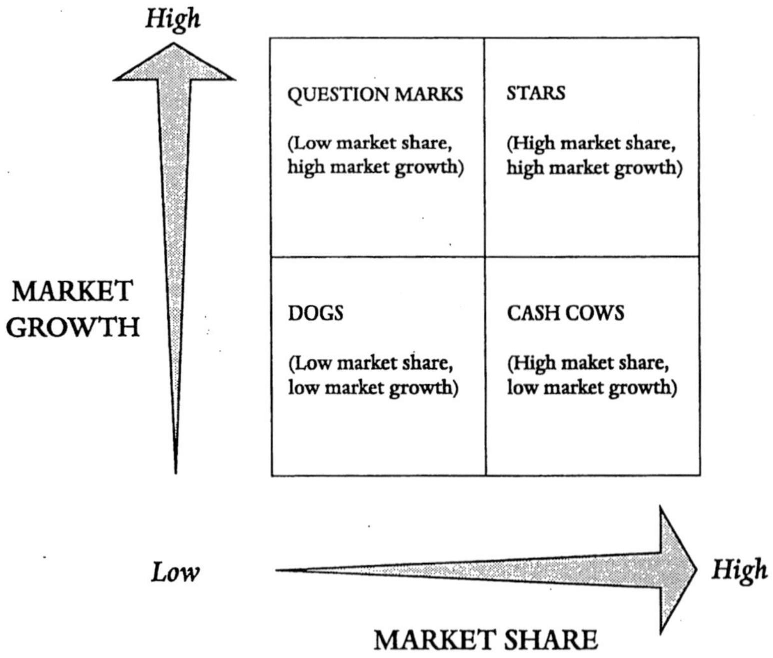

|  | Low market share | High market share |
|---|---|---|
| **High market growth** | QUESTION MARKS (Low market share, high market growth) | STARS (High market share, high market growth) |
| **Low market growth** | DOGS (Low market share, low market growth) | CASH COWS (High market share, low market growth) |

Vertical axis: MARKET GROWTH (Low to High)
Horizontal axis: MARKET SHARE (Low to High)

SOURCE: Barry Hedley, "Strategy and the Business Portfolio," *Long Range Planning*, February 1977: 12.

The internal dimension is market share. The assumption, in this case, is that for a business unit to have high market share, the production, finance, information systems, marketing, human resources, and research and development capabilities of that business unit must be above average. Again, only one decision has to be made. It must be decided whether the market share is high or low.

The resulting four matrix positions have each been given a name and a generic strategy. High-market-share, low-market-growth units or products are called cash cows. These are the most profitable ventures for a company, and help to generate strong cash flows. Investments should be made to sustain their cash-generating ability. High-market-share, high-market-growth units or products are called stars. They will become cash cows as market growth inevitably slows. While these ventures are profitable and generate cash, they also require investment to maintain their high market share as the market grows. Investments may be made in promotion, getting and keeping shelf facings, aggressive pricing (especially in the face of price wars) and product improvements through continued research and development.

Low-market-share, high-market-growth units or products are called question marks, and are, typically, new ventures for a company. It is always best to launch new ventures when the market is growing. As they are launched, they will, of course, have low market shares and will require substantial investment from the company so that awareness may be built through promotion, and wholesalers and distributors may be encouraged to carry the product. These new ventures will have already sustained heavy research-and-development and market launch costs, which will need to be recouped. Low-market-share, low-market-growth units or products are called dogs. If these are question marks that do not gain consumer acceptance before the market slows, it is probably best to sell off or close these ventures. If they are former cash cows which have lost their popularity due to changes in consumer needs or desires, or in technology, it is probably best to manage the ventures carefully until they are no longer contributing profits to the company.

The key problem with these two strategic approaches is that both a company and its competitors can perform the same analysis. It is hard to win a battle when your opponent knows exactly what moves you will be taking. Some of the most innovative marketing has been done by companies that did not follow the proposed strategy models. Carling Black Label beer was a dog, but some significant re-investment in the brand saw it resume cash-cow status for a few years. Most marketers favour more artistic models of strategy.

## ◊◊ Other Strategy Models

Harvard University's Michael Porter suggests there are five factors that need to be fully analysed before a strategy is developed (see fig. 5). One factor is the power of suppliers. If only one or two can supply the raw materials a producer needs, power rests with those suppliers, and the producer is made weaker. One also needs to assess the power of buyers or consumers. If hundreds or thousands of consumers purchase a company's products, power rests with that company. Will new companies be able to enter the marketplace? If there are many barriers to market entry, the opportunities for new competition are decreased, and power is gained by the established company. Are there products that can be easily substituted for those of the established company? The more substitute products that exist, the less power the company with the original product has. Finally,

*Figure 5.*   MICHAEL PORTER'S FIVE-FACTOR MODEL

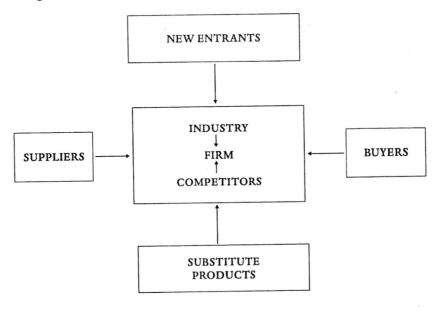

SOURCE: Michael E. Porter, *Competitive Strategy: Techniques for Analyzing Industries and Competitors* (The Free Press, 1980), 4.

one needs to assess the power of competitors. The more the dynamics of the marketplace are dictated by competitors, the less power a company has.

When a company develops a competitive strategy, it increases its power. Such a strategy works to raise barriers that prevent competitors from entering the market, to secure supply, to increase the number of potential consumers, to improve the company's position among competitors, or to make substitute products obsolete – but these actions cannot all be predicted from the strategy model alone. The additional insight of a manager is required, and that makes each strategy unique.

A different strategy model involves an analysis of the product/market focus of the company, both in the present and in the future. It is a useful brainstorming tool that can be used to generate and scrutinize the risks and rewards of future strategies (see fig. 6). Strategies can be designed for the existing target market to better serve its needs (a product may be relabelled or put in a new container) or for a new target market (a product may be launched in a different country, or its benefits may be adjusted to suit new demands). Strategies can also be designed to exploit existing products (a new scent to a laundry detergent or the product may be concentrated so less packaging is needed) or to develop new products (for example, Coca-Cola popsicles or Tide toothpaste).

*Figure 6.*   THE PRODUCT/MARKET FOCUS MODEL

|  | Old Product | New Product |
|---|---|---|
| **Old Market** | Product Improvement | Market Improvement |
| **New Market** | Product Diversification | Product Development |

SOURCE: Dale Beckman, David Kurtz, and Louis Boone, *Foundations of Marketing*, 5th ed. (Toronto: Dryden, Holt, Rinehart and Winston of Canada, 1992), 276.

The easiest strategies to implement, and those that pose the least risk, are in the upper left hand corner of figure 6. But serving existing markets with existing products will also result in the lowest additional rewards. Risk increases diagonally. The riskiest strategies are those that involve serving new markets with new products, but the rewards of such an approach can also be much greater.

A final strategy model, used extensively in introductory marketing courses, is shown in figure 7. The first step in S/W/O/T Analysis is to identify in detail the current strategy (the product, price, distribution, promotion, and target-market elements) for a company. Then the activities taking place beyond the control of the company, environmental factors, must be assessed. The concept of control is important here. Whatever a company can control becomes part of its resources as either a strength or a weakness. If a company can merely influence a factor, it is considered either an opportunity or threat in the environment. It is also important to evaluate the ethical standards of doing business; this has become an increasingly important concern.

*Figure 7.   S/W/O/T ANALYSIS*

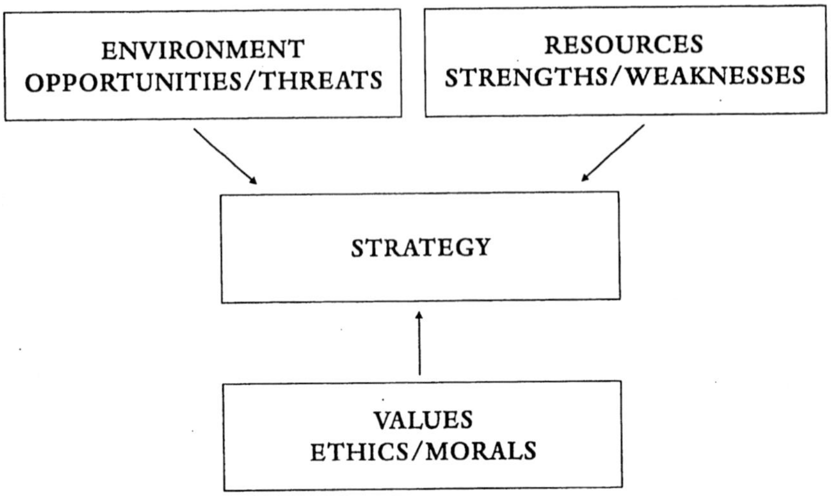

```
┌─────────────────────┐     ┌─────────────────────┐
│   ENVIRONMENT       │     │    RESOURCES        │
│ OPPORTUNITIES/THREATS│    │ STRENGTHS/WEAKNESSES│
└──────────┬──────────┘     └──────────┬──────────┘
           ↘                           ↙
              ┌──────────────────┐
              │    STRATEGY      │
              └────────▲─────────┘
                       │
              ┌────────┴─────────┐
              │     VALUES       │
              │  ETHICS/MORALS   │
              └──────────────────┘
```

SOURCE: Renato Mazzolini, "European Corporate Strategies," *Columbia Journal of World Business*, Spring 1975: 99.

The next step is deciding whether the current strategy can work in conjunction with the environmental and resource analysis just undertaken. If changes are necessary, they must fall within the ethical ground rules identified. For instance, the Sierra Leone government might decide to privatize a diamond mine. A certain company may have expertise in diamond mining, but, clearly, before company management moves to acquire the mine, it must consider the ethical and moral consequences of its decision.

In this model, strategies always start from a position of strength and are designed to: block environmental threats, take advantage of environmental opportunities, or correct resource weaknesses.

## ◊◊ A Word About Tactics

All of these strategy models might seem confusing, and no one model is right for every situation. Tactics are even more closely tailored to the situation at hand, and models are harder to find. Generally, a set of tactics is the implementation plan for the long-term strategy undertaken by a company. As a result, tactics involve the specification of a time frame, a set of prioritized actions, and control or check points, which are put in place to ensure that a given set of tactics is working. When formulating a tactical plan, one would never simply say to a company, "Do some advertising." One would specify the media to be used; the hours, weeks, or months of the campaign; and the type of message. In that tactical plan, one would list the most critical actions first, and the less important, optional actions second. A person would be unlikely to undertake a journey by car without specifying a goal, such as "I want to be in Moose Jaw by five o'clock tomorrow." In making a tactical plan, one must set or predict goals, and if these are not realized, some corrective action may be necessary.

When you work with case studies, you are developing your skills in specifying both strategic and tactical plans. Like all skills, these improve with practice. Case-study work will also teach you what famous strategists discovered years ago: drawing on the experience, perceptions, and viewpoints of your peers will improve your own strategic perspective.

*Marketing Insights*

# MARKET SEGMENTATION

Many children are familiar with Saturday-morning television commercials for breakfast cereals such as Lucky Charms and Count Chocula. The brightly coloured animated characters that populate these commercials capture children's imagination, and the promise of special prizes inside the boxes of these sugary cereals makes their appeal almost irresistible. But did the manufacturers of those cereals make a mistake?

While children are more than willing to purchase their products, most do not buy them. The actual cereal purchasers are parents. There is no doubt children have an influence, but their parents make the actual purchase decisions. And their decisions aren't always correct. After buying a box of strawberry-flavoured cereal, a parent, rather than his or her children, may have to finish the remainder of the box.

Today, manufacturers understand that a cereal has to appeal both to the child and to the parent. Those who sell cereals must provide lists of ingredients, notations about added vitamins and minerals, labels highlighting all natural ingredients, and mentions of reduced sugar and salt. There are still special toy surprises and animated characters to encourage children to influence their parents, but the role of children in the purchase process is much better understood.

## ◊◊ The Marketplace

While we think of a marketplace as being a location, it is more correctly defined as people. Manufacturers do not sell goods to places; they sell them to individuals. This is true of both consumer goods, which are sold to individuals for their own personal use, and industrial goods, which are products used directly or indirectly in the production of other goods or for resale. It is people making decisions for themselves or for their company who buy products.

But the marketplace is not just people. If it were, companies would be focusing on the most highly populated places in the world – China and India. But they don't. A market also requires a willingness to buy. The introductory example of breakfast cereals shows that willingness to buy is a necessary, but not in itself sufficient, criterion for a market to exist. Children are willing to buy many things, yet they do not make many purchases. To have a true marketplace, we need people who also have purchasing power and authority. While China and India have, together, nearly three billion people, the per-capita income or purchasing power for these people is so low that they cannot afford to purchase many of the standard household goods we take for granted in North America. Authority to buy becomes a significant issue in industrial markets. A company trying to sell a new line of personal computers to a university can waste much time talking to individual faculty and staff members who do not have any authority to sign a purchase requisition or place an order.

Remember our definition of marketing: the art of finding out what people want and giving it to them in a way better than the competition while generating revenue over the long term. Even when a market can be identified, it probably contains so many people that determining and satisfying all their needs is impossible. For instance, ask the people in your class what they want in an automobile and you will receive a gamut of responses. Trying to build the one automobile that will satisfy those heterogeneous desires is impossible. One would be wiser to find groups of people within the market who have similar desires and build an automobile to satisfy them. It is nearly impossible to satisfy all of the people, all of the time with one product, so marketers try to satisfy some of the people, all of the time. Even manufacturers of food staples such as sugar and flour have recognized that their products are consumed by several different groups in the marketplace and have developed product variations to increase customer satisfaction.

This process of dividing the heterogeneous marketplace into smaller, more homogeneous chunks is called market segmentation. Once the market is segmented, a marketer can develop products and a marketing plan that will better meet the needs of the people who make up each segment and improve the chances of making a sale. The ways in which the marketplace can be segmented vary for consumer and industrial products.

## ◊◊ Bases of Consumer Market Segmentation

There are four bases for segmenting the market for consumer products (see fig. 1). The oldest basis is segmentation by geography. In Canada, a little over sixty-one percent of the population live in Ontario and Quebec. Cities account for eighty percent of Canada's population. The country's three largest cities – Toronto (6.1 million), Montreal (4.1 million) and Vancouver (2.5 million) – account for thirty-five percent of its population. Climate is also a basis for geographic segmentation. People in Vancouver rarely see snow, and people in Regina have more sunny days than those in any other provincial capital. If one is looking for a suitable location for a professional-sports franchise, it is important to consider gross city size. Many believe that Saskatoon and Hamilton are too small to support an NHL hockey team. If one is considering selling products on a door-to-door basis, population density is an important factor. Rural areas, with low population densities, make selling door-to-door very impractical.

As the mass media grew in importance during the late 1800's and early 1900's, a second basis of segmentation was born: demographics. Age, gender, income, occupation, education, whether a person's home is owned or rented, number of children, and ethnicity are some demographic variables. As an aid to advertisers in deciding which medium to use, companies began to gather and share this type of information. Demographics are easy to gather and common sense dictates that they are associated with the purchase of many products. Diapers are consumed mostly by people with children under the age of two. High-heeled shoes are consumed mostly by women. University textbooks are consumed mostly by people aged eighteen to twenty-two. Pasta is consumed twice as often by people of Italian heritage.

Figure 1. FOUR BASES OF CONSUMER MARKET SEGMENTATION

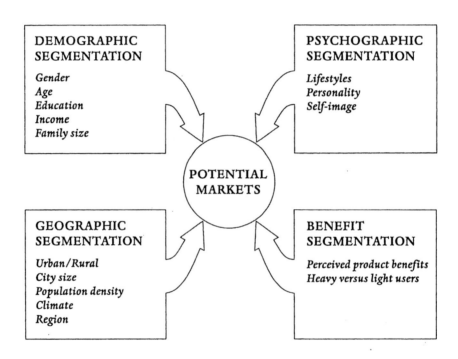

SOURCE: Adapted from Jack Z. Sissors, "What Is a Market?" *Journal of Marketing*, July 1966: 21.

You should know a few current demographic trends. Baby boomers, those born between 1946 and 1964, are getting older (see fig. 2). In North America, the segment of people over the age of fifty is expected to show double-digit growth over the next decade – a much higher rate of growth than that of any other segment. Compared to twenty years ago, people are marrying at a later age and they are having fewer children. More than one in two marriages will end in divorce. People are saving money for their retirement, especially those with higher incomes. In fact most of the disposable wealth in Canada is controlled by those over the age of fifty. Compared to twenty years ago, people are also more educated and more are employed in white-collar jobs.

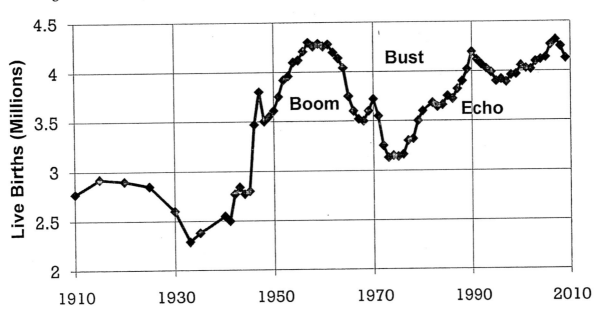

Figure 2.   ANNUAL NUMBER OF BIRTHS IN THE UNITED STATES

Demographics, however, are not as good a predictor as they were a generation ago. Television, the other mass media, and social media are partly to blame for this. Today, there are people over the age of sixty running triathlons, while many people under the age of thirty are out of shape. Some people with low incomes drive expensive luxury cars while some people with high incomes drive economy cars. As marketers have discovered, an increasingly important basis of segmenting the market is psychology. Psychographics attempts to blend demographic data with psychological factors, such as attitude and lifestyle, to develop a behavioural profile of consumers.  As a result, we find that some consumers are old-fashioned while others are adventurous, and that some groups are family-centred and others live for the moment. Each of these groups will buy different products to satisfy their emotional needs.

A final basis of segmentation is by expected product benefits. While you and I may each purchase a box of baking soda, you may be buying it to deodorize your refrigerator while I plan to do some baking. Take this test. What brand of toothpaste did you use this morning? Chances are that if you used Crest or Colgate, you wanted cavity prevention and plaque protection. If you bought Close-up or Pearl Drops Tooth Polish, you wanted a sexy, white smile. If you bought Aquafresh, you wanted triple protection using its three stripes: fresh breath, cavity prevention, and white teeth. People buying Macleans or Pepsodent wanted to save some money. One might also segment the market by the volume consumed. An old rule of thumb is that 20-percent of a company's customers account for 80-percent of the volume sold. There are people who consume five cups of coffee per day. Companies do not want to offend these people with any new marketing activity. Furthermore, companies like to identify these consumers to determine how lighter users can be turned on to the product and be made heavy users.

## ◊◊ Bases of Industrial Market Segmentation

There are three ways to segment the industrial market (see fig. 3). As it can in the consumer market, geography can be used. In Canada, nearly three-quarters of the industrial market is located in a narrow band running from Windsor, Ontario through Toronto and Montreal and ending in Quebec City. While there are over thirty-five million consumers, there are only several hundred thousand industrial customers. Often industrial customers cluster around geographic features. For instance, deep-sea fishing firms are concentrated along the east and west coasts, while mining concerns are concentrated in the Canadian Shield regions of Ontario and Quebec.

A consideration of the end use of a product can help marketers to segment the industrial market. Consider the industrial market for personal computers. Firms may purchase personal computers to act as print servers, stand-alone workstations, monitoring workstations for industrial processes, database servers, desktop-publishing centres, or as hubs of e-mail networks. While the computer required to work with each of these applications is essentially the same, the marketing plan developed by a computer company to reach each industrial customer is different.

Finally, the need for special products can be used to segment the industrial market. A valve that can be opened and closed to permit or cut off the flow of liquid might, at first, seem to be a standard product. However, different industries have needs for valves with different properties. The transport of pressurized gases requires a valve that can operate at subzero temperatures. In the transportation of drinking water, valves need to be quite large to accommodate the volume of water moved in a city. In a chemical plant, valves will need to be designed to withstand chemical reaction. For an industrial marketer, it might be convenient to segment companies on the basis of their unique product needs.

Figure 3.   THREE BASES OF INDUSTRIAL MARKET SEGMENTATION

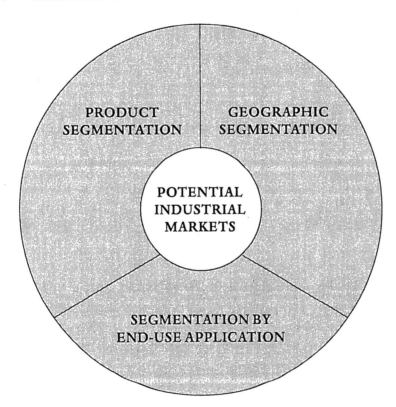

SOURCE: Dale Beckman, David Kurtz, and Louis Boone, *Foundations of Marketing*, 5th ed. (Toronto: Dryden, Holt, Rinehart and Winston of Canada, 1992), 75.

## ◊◊ Market-Segmentation Strategies

How should a company use this knowledge of market segmentation? There are three potential strategies that can be used. The first is to treat the entire market as an undifferentiated whole and offer it a single product.

Companies taking this approach admit that they cannot satisfy all of the people all of the time, yet they try to satisfy most of the people, most of the time. Coca-Cola was invented just over 125 years ago. For the first eighty years of its life, it produced only one kind of Coke - with sugar and caffeine. Such a single-offer strategy allows for economies of production and savings on promotion and inventory costs, yet reduces satisfaction among consumers.

A second strategy is to recognize that there are many different consumer segments and to develop a product for each segment's special needs. The history of marketing has shown us that after two brands have competed head to head using an undifferentiated strategy, it is most often the number two brand which sees the differentiated strategy as an opportunity to wrest the lead from its competitor. In the cola battles, it was Pepsi-Cola which first introduced artificially sweetened Diet Pepsi, caffeine-reduced Pepsi Free, and eliminated artificial colours in Crystal Pepsi. While production efficiencies declined and promotion and inventory costs climbed, Pepsi was rewarded with increased sales, increased consumer satisfaction, and increased profits. Not surprisingly, Coca-Cola quickly abandoned its old, undifferentiated strategy and followed suit.

A final strategy is, again, to recognize that there are many different consumer segments, and to simultaneously recognize that a company may not have the financial, production, or marketing resources to pursue a multi-offer strategy. Instead, it may choose one market segment and develop a product especially for it. In the soft-drink market, Orange Crush, Dr. Pepper, Canada Dry, Schweppes, Red Bull, and Hires Root Beer are all examples of brands that have followed a concentrated or niching strategy. The only trick to choosing a market niche is to make sure the niche is large enough to justify pursuing it, and that it is stable enough to endure. For a brief time in the 1970's and 1990's, platform shoes made a fashion statement. However, by the time some companies had developed their line of platform shoes, the fashion trend had passed on.

# PRODUCT STRATEGY

In 1993, Lee Iacocca retired as Chairman of the Chrysler Corporation. By the time Iacocca had become President in the early 1980's, Chrysler had suffered major losses and was rapidly losing market share. Though Chrysler was the third largest car company in America, most analysts predicted that Iacocca's appointment was too little too late to save the company. Stock prices had fallen and commercial debt had been so devalued that someone buying a bond could expect a 40-percent return, assuming the company survived.

By 1993, Chrysler's market share had recovered, though the company remained in third place, and its quarterly profits exceeded those of both its competitors. Iacocca has written several books on his successful turnaround of Chrysler Corporation. In these, he attributes his success to several factors, but much of it can, in fact, be attributed to his product strategy.

Chrysler was the first American car company to offer an extended warranty plan. Its seven-year, seventy-thousand mile warranty on the automobile powertrain nearly doubled the best American car warranty at the time, and set a new standard that its competitors had not managed to emulate fully by 1993. Chrysler was the first American company to introduce front wheel drive vehicles. It was the first to develop an automobile resembling a crossbred van and station wagon: the mini-van. Chrysler successfully reintroduced the convertible, which had disappeared in the early 1970's. Through its acquisition of Jeep, Chrysler became the first company to introduce "shift-on-the-fly" two-wheel-drive-to-four-wheel-drive vehicles. Chrysler was the first American company to make air bags a standard feature of its cars. As a parting salvo, Iacocca introduced cab-forward design. By moving the wheels of the automobile closer to the front and back corners, and by extending the passenger compartment, the company created a car that was both more spacious and more stable on the highway.

Lee Iacocca recognized that a company's product strategy goes beyond the physical properties of the products it sells. The total product strategy incorporates brand names, trademarks, labelling, packaging, consumer credit plans, warranties or guarantees, and brand image. It also includes service – the intangible aspects of a purchase.

## ◊◊ Products and Services

In 1945, half of all people who worked did so in the manufacturing sector. Today, people who make products account for only 15% of the workforce. Half of all people who work are now employed in the services sector. Services are different from products in several important ways. First, services are intangible. You cannot see, feel, or sample a service before it is performed. Services are difficult to standardize. Even if the same person was to cut your hair, no two visits to the hairstylist would yield exactly the same result. Services cannot be produced in advance and inventoried; idle capacity is, therefore, an important concern. We design ambulance services to have plenty of idle capacity, but dentists would go out of business without a full schedule of appointments. Finally, most services are produced and consumed simultaneously. When you travel by plane, the type of flight you experience depends as much on the other passengers aboard and the weather you encounter as it does on the flight attendants and cockpit crew.

Both products and services can be classified by the amount of effort they require of consumers. Convenience products and services are purchased frequently, as soon as they are needed, and with a minimum of effort. These might include bread, milk, potato chips, gasoline, dry cleaning, and fast-food meals. Shopping products and services are purchased only after the consumer has made comparisons between brands or stores. Consumers expect the monetary savings or greater satisfaction to outweigh the costs of shopping. Shopping products and services might include clothing, furniture, appliances, auto repairs and insurance. Specialty products and services possess some unique characteristics that cause consumers to prize them, and go out of their way to buy them. These might include services of specialized surgeons and divorce attorneys, or imported crystal, handmade sweaters, and classic automobiles.

## ⬥⬥ The Product Life Cycle

A key concept is the product life cycle. This represents the typical path a product or service follows from introduction to deletion (see fig. 1).

In the introduction phase, a company builds awareness of, and stimulates demand for, the new product. Promotional campaigns stress information about the product to educate the general public. Wholesalers and retailers are convinced to carry the new product and to give it shelf space. Continuous research and development is undertaken to refine the product to match consumer expectations. Though pricing policies can be adopted that increase returns, losses are expected during this phase because of significant promotion and research costs.

In the growth phase, sales rise dramatically as consumers first try, and then accept, the product. As profits grow, so does the number of competitors, as introducing a new brand of a product is easiest when the market is growing. Promotional campaigns begin to stress corporate brand names and consumers are induced to make trial purchases.

Mass-media advertising is used to reach consumers and to stimulate publicity. It becomes easier for a company to add more wholesalers and retailers to its lists as its product develops a sales track record. In the late growth phase, prices begin to decline as competition increases. Research and development remains intensive as producers search for the consumer ideal.

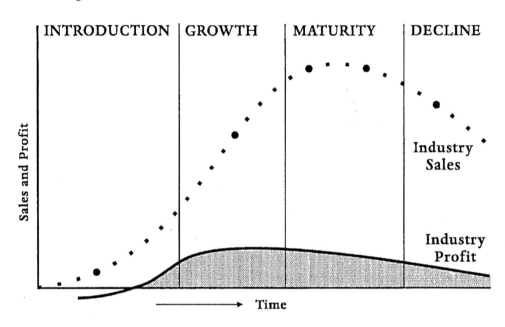

Figure 1. PHASES IN THE PRODUCT LIFE CYCLE

SOURCE: Dale Beckman, David Kurtz, and Louis Boone, *Foundations of Marketing*, 5th ed. (Toronto: Dryden, Holt, Rinehart and Winston of Canada, 1992), 243.

In the maturity phase, sales reach a plateau and competition is at a maximum. This inevitably leads a company to a state of overcapacity and to make cutthroat competitive moves in order to hold market share and force weaker companies out of the market. Prices continue to decline, and price wars often occur. Profitability of the entire industry suffers. Those who have done the research-and-development work have identified customer ideals so well that products are virtually indistinguishable. Promotion emphasis shifts to retaining customers and building repeat purchases. Brand names and subtle product differences are stressed in expensive mass media campaigns. The number of wholesalers and retailers carrying the product reaches a maximum during this phase. As competitors are forced out of the market, some wholesalers and retailers begin to drop the line from their product mix.

The decline phase is characterized by a permanent drop in industry sales caused either by new technological innovations or by changes in basic consumer preferences. Competition for fewer sales means more overcapacity and more companies leaving the market. Industry profits continue to fall, and may become losses. No money is spent on research and development – products are not improved. Because there are fewer competitors, prices begin to stabilize and may increase slightly. The number of distributors continues to fall. Promotion is cut back to minimal levels and messages are more reminder-oriented.

The time it takes for a product to move through its life cycle can vary dramatically. The incandescent light bulb was invented more than 140 years ago by Thomas Edison, and is just entering the early decline phase of its life cycle. The black and white television set was invented in the early 1940's and today it is in the late decline phase – about to disappear. Laser disc players had a complete life span of less than a decade in the late 1990's and early 2000's.

## ◊◊ Exploiting the Product Life Cycle

A single-product company lives and dies by the product life cycle. Most company managers have a portfolio of products at different stages of their life cycles. One should also realize that, if they are properly managed, products in the decline phase of the product life cycle can still be profitable.

The decline phase does not have to follow the maturity phase if sales of the product can be kick-started (see fig. 2). There are three ways to extend the product life cycle: by finding new uses, by finding new users, or by increasing usage among current users. Baking soda faced a decline, as people did not have the time or inclination to continue home baking. To keep the product alive, many new uses for baking soda were devised; it could serve as a freezer or refrigerator deodorant, cat-box freshener, bathwater additive, toothpaste whitening ingredient, rug cleaner, breath freshener, or household cleanser.

Soft drinks are mostly consumed in much greater quantities in the summer, when a cold beverage is most welcome. In the winter months, people turn to coffee, tea, and hot chocolate. In an effort to rejuvenate soft-drink sales, many companies are experimenting with hot versions of their product to appeal to a new group of consumers. Just imagine a steaming cup of cola on a cold winter's day.

*Figure 2.* EXTENDING THE PRODUCT LIFE CYCLE

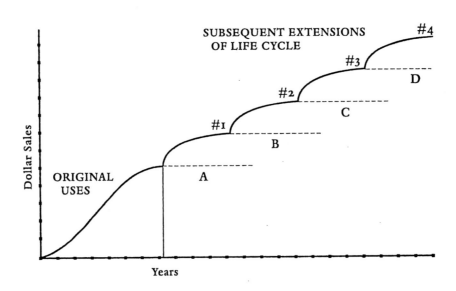

SOURCE: Theodore Levitt, "Exploit the Product Life Cycle," *Harvard Business Review*, November–December, 1965: 88.

As a rule of thumb, 20-percent of customers account for 80-percent of a product's sales. If a company could transform more light users into heavy users, it could extend the product life cycle. Some companies have added a second target market to add new users. A Canadian company could start selling in China or India. A company targeting men could target women as well.

To increase usage, one could change the package size. In Canada, twenty years ago, soft drink companies changed their can size from 280 to 355 millilitres, thus triggering a 25-percent increase in sales volume. Another ploy would be to have consumers save many box-tops for a limited time or they could collect punches on a card each time they purchased a cup of coffee. Consumers would thereby be encouraged to increase their product usage if they wanted the bonus prize.

Some texts suggest a fourth way to rejuvenate a product's life cycle: by reformulating the product, changing its quality, using a new package, or changing the label. However, these are simply tactics to increase usage, attract new users or find new uses.

## ◊◊ Developing New Products

To fully exploit the product life cycle, companies must always be seeking new products. A company might use internal research-and-development projects, or it could find externally generated technologies that lead to new products. A product is considered new when it involves an area of activity or a production process new to the company. Historically, developing successful new products has not been easy. Studies conducted in the 1960's, showed that two in three new products were a commercial failure. Academic researchers and private corporations have tried to determine why new-product development is not more successful.

As we see in figure 3, all new-product projects must move through six phases of development. At any phase, a project idea might be dropped. The first phase is idea generation. This is the least costly of the six phases, yet it can be the most difficult. People who are asked to brainstorm new-product ideas need to set aside their critical judgement and be free to create. Often the first idea generated is not practical, but that idea may lead to another idea, and yet another, which then can be commercially developed. The key is not to dismiss any idea as being too crazy. As an exercise, examine a belt. Focusing on the belt itself or on the process by which it was produced, try to think of other products that could be developed. Those people who can "ideate" freely are able to generate more than thirty new product ideas in fifteen minutes or less.

The following phases require an increasing commitment of company resources to the new-product project. A company must carefully screen this creative list of new-product ideas. Inferior ideas must be filtered out in such a way that potential winners are not inadvertently dropped. Screening involves a quick assessment of the five environments (technology, competition, political/legal, economic, and socio-cultural) as well as the company's strengths and weaknesses. Ideas that pass this initial screen are developed into full business plans. Detailed assessments of competitors, spreadsheets showing estimated capital expenditures and operating income, and market research into consumer attitudes are assembled. Rigorous quantitative hurdles are set for these business plans so that company resources are invested in projects that will increase shareholder wealth.

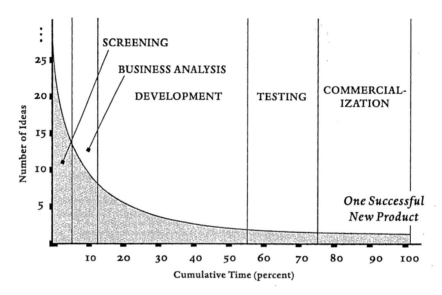

Figure 3. DECAY CURVE OF NEW-PRODUCT IDEAS

SOURCE: Adapted from *Management of New Products* (New York: Booz, Allen & Hamilton, 1968), 9.

Those plans that are approved lead to product development. At this stage, product prototypes are developed and tested with actual consumers. In the industrial market, new process technologies are tried in pilot plants to ascertain optimal operating conditions and guarantee quality of output. Forty years ago, the Carnation Milk Company discovered a market opportunity for chocolate skim milk powder. The new-product project died in the product development stage, as no formulation could be developed to meet the government standards for fat and calorie content in skim milk.

If a company is certain about the viability of a product and its marketing plans, it might move straight to commercialization – making the product available in the national marketplace. Sometimes, companies have some details to fine tune. A test market – an optional stage – could be undertaken at this point. To conduct this test, marketers choose a small city that is representative of the target population. The product is introduced here, and a small-scale replica of the marketing plan is reproduced – down to the level of couponing, advertising and store displays. The results are closely monitored and activities adjusted to find the optimal marketing plan.

While test markets can cost a half million dollars each, they can be cheap insurance to prevent problems that can arise in a multi-million-dollar national product launch. McDonald's Restaurants has used test marketing successfully to gauge reaction to salads, pizza, chicken nuggets, wraps, and a ground-pork sandwich. However, a test market can reveal a company's plans to its competitors. Colgate has been fighting for nearly five decades to regain, from Crest, its position as the number-one-selling toothpaste. For adults, the number-one dental problem is not cavities but gum disease. Colgate developed an adult toothpaste formulation and used test marketing to gauge reactions. While the new formulation was a success in test market, a short time after Colgate introduced it nationally, Crest introduced its own version of the toothpaste. Crest had learned of the innovation during the test market. Colgate was unable to gain a sustained competitive advantage.

Some companies try to foil a competitor's test market by changing their own marketing activities. For instance, they may distribute extra-value coupons or free samples, temporarily lower prices, or increase their volume of advertising to make the competitor's test market seem unsuccessful. The blockage of potentially successful new products is part of the free-enterprise system.

To increase the success rate of new product commercialization, Dr. Robert Cooper and Dr. Elko Kleinschmidt of McMaster University studied successful and failed product launches. They revised the old model slightly (see fig. 4). They proposed that an effective new-product process was a series of stages and gates. The gates represent hurdles to be overcome. Their research indicates that new products fail because companies don't do all the groundwork required in a given stage, and that hurdles or gates, at which go/no-go decisions are made, are sometimes skipped or hurdles lowered. They advocate for a rigorous approach to new-product development in which full research is conducted with the consumer at every stage. After four decades, their research is paying dividends. Their approach has been adopted by most of the Fortune 500 with those companies seeing two in three commercially launched new products succeeding.

*Figure 4.*   A STAGE-GATE APPROACH TO NEW-PRODUCT DEVELOPMENT

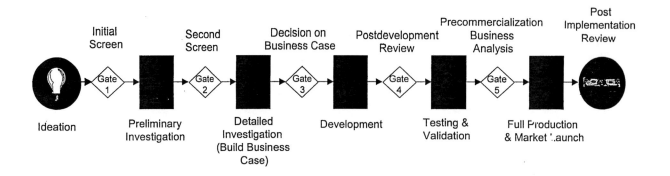

SOURCE: *Winning at New Products* (Toronto: Addison Wesley, 1991), 73.

## ◇◇ Product Identification

Many products are in the maturity phase of their life cycles, and it is their brand images that affect a consumer's willingness to purchase them. Brand image can be conveyed through the use of logos or symbols, brand names, and packaging.  As trade becomes more globalized, the use of logos or symbols will increase, because these devices do not need to be translated from one language to another.  Witness the success of McDonald's golden arches or Disney's Mickey Mouse.

The average university graduate has a speaking vocabulary of eight thousand words, yet the average supermarket has over ten thousand individually branded products.  Certainly, not all brand names will be equally effective. What makes a brand name good? First, it should be easy to pronounce, recognize and remember.  Most effective brand names have one or two syllables and just a few letters: Tide, Bold, Cheer, Dial, Joy, Dove, Kraft, Coke, Pepsi, Jell-o, Blue and Canadian. A study of trademark files indicates that almost all four- and five-letter words have been used in North America as brand names. Some companies are legally protecting six- and seven-letter words for that future time when they will need to introduce new products.

Second, the name should give the right connotation to the buyer. *Connotation* refers to a suggested or implied meaning. *Denotation*, however, refers to the exact meaning of a word – its dictionary definition. *Tubular* means "pertaining to, shaped like, or consisting of, one or more hollow cylinders, or tubes." When a Teenage Mutant Ninja Turtle says, "It's like tubular, dude," the word has a connotative, not a denotative, meaning: it connotes "special," "excellent," "radical." The name given to a product should always have the correct connotation. "Slender" is a great name for a diet drink; "Gaunt" is not.

Finally, the name should be capable of being legally protectable. In this day of designer waters with foreign names, expect to see Perrier, Evian, and Montclair facing competition from L'Eau. But the latter is not a brand name that can be protected. You cannot force French-speaking people to stop using the word *eau* ("water") in their advertising. This inability to legally protect generic terms has come to haunt some companies. Twenty years ago, a court ruling prevented Labatt Breweries from protecting the name Blue, and, by the same token, Molson Breweries cannot protect the name Canadian. Some companies fight to remind consumers that their brand name is not the same as the products they represent. Scotch Tape is an adhesive tape. Aspirin is an acetyl-salicylic-acid pain reliever. Kleenex is a facial tissue and Jell-o is a gelatin dessert. If the fight is lost, these names may be declared unprotected, just like escalator and zipper, former brand names that are now free to be used by any company.

Firms can choose either a family-brand or an individual-brand approach. Kraft is a brand name for a family of food products: cheese slices, peanut butter, cream cheese, salad dressings, mayonnaise, and margarine. Family brands allow for an economy of promotion. Money spent to advertise any one product benefits, to some degree, all the other products in the line. Also, consumers who have a positive experience with one item in the product line often make trial purchases of other items in that line.

In the case of individual branding, each item in a product line receives a unique brand name. You might be hard pressed to name all the products manufactured by Procter and Gamble because the Procter and Gamble name is not made prominent on most of the company's packaging. Individual brands allow for better targeting of products to specific (and probably different) markets. A company that employs individual brands can also have more than one brand competing in the same product class. Procter and Gamble manufactures three different laundry detergents, and they compete against each other. While individual brands are more expensive, companies hope that improved customer satisfaction will increase revenues.

Packaging and labels also define a brand image while serving many other functions. Clearly, packaging protects the contents from damage and spoilage. It helps reduce shoplifting. Small items are placed in larger boxes to prevent the item from being slipped up a shoplifter's sleeve or down his or her boot. Packaging allows consumers to re-use a product. Cereal boxes are designed to store the product between servings, and to keep it fresh over time. In a retail store, package fronts act as small billboards. Consumers may have no idea which brand they will purchase until confronted by the packages in the store.

Labels contain information in both of Canada's official languages. Examine a package of cigarettes, a can of soda or a bag of potato chips, and you will discover more than the brand name or company logo. You should find the name and address of the manufacturer and/or distributor, information about the product composition and the package size, the weight of the contents, recommended uses of the product, promotional information, and the universal product code. The latter has been designed to speed consumers through checkouts, to reduce salesclerk errors and to improve inventory control.

# PRICING STRATEGY

A child opens a package of hockey cards. "Look Mom," the child exclaims, "it's a Sidney Crosby rookie card. I bet it's worth five hundred dollars." Mom doesn't look nearly as excited as the child. Perhaps it's because Mom knows that she couldn't take the hockey card to her bank and use it to reduce her mortgage payment by five hundred dollars.

In the classic sense, price has often been defined as the exchange value of a product or service. The definition implies that something is only worth what someone else is willing to trade for it. Perhaps someone might have been willing to trade five hundred dollars for the hockey card, but it is doubtful that a plumber would take the card in exchange for services rendered.

From time to time, especially when the economy is in a recession, the practice of bartering generates renewed interest. In those unique times, a few individuals revive the concept of exchanging one product or service for another. However, most companies deal in cash dollars, so price represents the cash amount a person or a company is willing to exchange for a product or service.

Producers of products or services must make three classic business decisions: what price level will be set; what discounts will be allowed to purchasers; and what impact the first two decisions will have on the profitability of the business. Let's examine each of these questions in a little more detail.

## ◊◊ Setting a Price

Classic microeconomic theory suggests that a company needs to determine two curves. One represents the quantity that consumers will demand at various price points. Its shape depends very much on the competitive circumstances in the market (that is, whether a company has a monopoly or whether it faces some degree of competition). The second curve represents the cost a company faces in producing various quantities of a product. Using calculus, one can take the first derivative of each curve to give both a marginal-cost and marginal-demand curve. By finding the intersection of the marginal curves, one has found the point at which the last dollar of cost has been balanced by the last dollar of revenue. Having made this determination, a company would then price the product at that point.

There are two major problems with that approach. The first concerns the construction of the curves. Estimating a reliable cost curve for a company is a fairly easy task. The same is not true of a demand curve. Even if one could reliably determine a demand curve at 9:00 a.m., because consumer tastes and wants are constantly in flux, that demand curve would be out of date by 10:00 a.m. Conceivably, if one used microeconomic theory, the price of a good could change in the time it took for a consumer to pick it from the shelf and pay for it at the checkout counter. As we need an easier and more stable method of pricing, microeconomic theory is generally rejected.

The typical approach to pricing is a cost-plus approach. With the help of its accounting department, a company can determine all the costs of producing a product. There are two types of cost: fixed costs, which do not vary much with the quantity produced; and variable costs, which vary directly with each unit produced. Typical fixed costs include administrative salaries, heating and electricity, property taxes, rent, and promotion budget. Variable costs include raw materials, component parts, and assembly labour. At a minimum, the price paid by consumers must cover all variable costs and should include an extra amount to cover the fixed costs and a contribution to profits.

This extra amount is called the mark-up. While many marketing texts like to define mark-up based on price, it is easier to think of it as based on cost. For instance, if the variable costs of a product are three dollars and a company chooses to sell the product for four dollars, the mark-up of one dollar could also be expressed as a 33.3 percent mark-up based on cost.

The classic business dilemma here is just how big a mark-up to take. Clearly a mark-up which is too big will not be well-received by consumers, and the product won't sell. A mark-up that is too small will be well-received by consumers, but the sales won't generate enough money to cover the fixed costs of doing business. This dilemma is further complicated when a producer realizes that marketing intermediaries, such as wholesalers and retailers, will also mark-up a product to cover their costs. Determining the price at which to sell a product to wholesalers while also calculating a suggested retail selling price is something of an art (see fig. 1).

The size of a mark-up is generally related to stock turnover. The more frequently an item is sold, the less money has to be made on each sale to cover costs. Milk is a product that is sold frequently. Typically, a grocery store's inventory of milk is completely sold in forty-eight to seventy-two hours. Given this high turnover rate, the mark-up on milk is typically 1 or 2 percent. In that same grocery store, one may find dishware or glassware. This inventory will be completely sold in four to six months. Given this low turnover rate, the mark-up on these products could be as much as 100 percent. The average grocery store mark-up is probably 5 to 8 percent. In another industry – for example, retail sporting goods – the average is probably 100 percent, with some items marked up as high as 300 percent.

MARKETING INSIGHTS

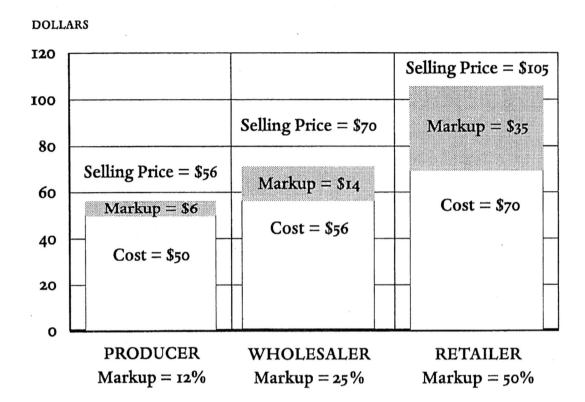

Figure 1. THE MARK-UP CHAIN

◊◊ A Look at Discounts, Allowances and Rebates

Although a product is assigned a suggested selling price, the price a company or consumer actually pays for it can be quite different. This market price might reflect various discounts or additional charges.

Discounts are reductions in the price one pays. The need to move inventory is one reason that companies apply discounts to their products. For example, a tomato-juice manufacturer could still have inventory on hand when the new harvest of tomatoes begins. The company needs to clear the warehouse before the new production run, and so it might consider a quantity discount. If a typical case price to the wholesaler is twelve dollars, it could be reduced to ten dollars, but only if the wholesaler agrees to purchase a minimum of one thousand cases.

Another reason a company might offer a discount is to encourage consumers to pay their bills quickly. The company can thus generate a larger or more liquid cash flow. In recessionary times, making a sale may not be enough. Prompt settling of accounts receivable helps to generate the cash lifeblood of a company. In purchasing five thousand dollars worth of lumber from a building-supply store, a contractor might be offered a 2 percent discount if he or she is willing to pay in cash rather than being issued a bill payable in thirty days.

Finally, a discount may be given in exchange for marketing functions that might normally be performed by the manufacturer. Sometimes wholesalers or retailers perform special marketing services for a manufacturer. For instance, a retailer might allow a manufacturer to display its product in an end-of-aisle display, or to set up an in-store tasting. Likewise, a wholesaler may agree with the manufacturer to implement a new shipping and storing arrangement using a new pallet size. In these cases, manufacturers offer discounts to thank and reward intermediaries for their efforts.

The three discounts discussed may be pass passed along to consumers in their entirety, or in part, or not at all.

Allowances should be considered separately. These do not change the price paid for an item, but represent credits that can be applied to a purchase. Consumers are well aware of trade-in allowances. When purchasing a new car, a consumer will negotiate with the car retailer to obtain the best price possible. With that deal completed, the consumer might offer to sell the car he or she is currently using to the dealer. A new negotiating process ensues, and a final trade-in allowance is determined. The allowance does not change the negotiated price for the new car, but represents a credit that can be applied to the purchase price. When purchasing a new car, you may not like the trade-in allowance offered by the dealer. So, instead, you may try to sell your used car privately hoping to get more than the trade-in allowance.

Another type of allowance is for promotion of the manufacturer along with the retailer. Every fall, car manufacturers must unveil all their new-car models for the coming year. Car retailers are often given promotional allowances to organize unveiling events that promote both the dealership and the new cars. You might know dealerships that have hosted barbecues, installed special lighting, or temporarily erected tents to announce their new models. These were all financed by allowances.

When the economy slips into a recession or depression, it is not uncommon to see the rebirth of the rebate. A rebate is a refund by the seller of a portion of the purchase price. A sure sign of a slipping economy is the renaissance of rebates on new-car purchases; automobile manufacturers have used this technique extensively. It is also possible to have smaller rebates on the purchase price of household appliances.

## ◊◊ Shipping Charges

A final adjustment to the market price involves freight or shipping charges. One can imagine two extremes here. The first would be that the purchaser pays all shipping charges. This is often quoted as free-on-board (or FOB) plant pricing, and suggests that the price quoted is applicable at the plant gates. Should the purchaser want the product delivered anywhere else, it is responsible for all shipping charges.

The other extreme would be a delivered price. This is called freight-absorption, and refers to the price of goods when they have been delivered to wherever the purchaser specifies. A smart purchaser will ask for quotations of both the FOB plant price and the freight-absorption price. The difference represents the cost of freight for the seller. If the purchaser can arrange delivery more cheaply, it will accept the FOB plant price. If it cannot, it will accept the freight-absorption quotation.

Calculating the actual freight charge for each and every order would be a time-consuming task. Instead, a company can calculate an average delivery charge by adding all delivery charges over the previous twelve months and dividing that total by the total number of products shipped. The result is called a uniform-delivered price. Unfortunately, employing this pricing strategy penalizes those who are located close by; they must subsidize the freight costs for those who are situated farther away. The Canadian post office, for example, charges the same price to deliver a letter across the street as it does to deliver a letter to an address several provinces away.

To modify the uniform-delivered pricing plan, one could create concentric zones around the seller's location. Within a zone, there is a uniform-delivered price, but that price increases when one moves from a zone closer to the seller to a farther one. This is called zone-delivered pricing (see fig. 2).

*Figure 2.* ZONE-DELIVERED PRICING

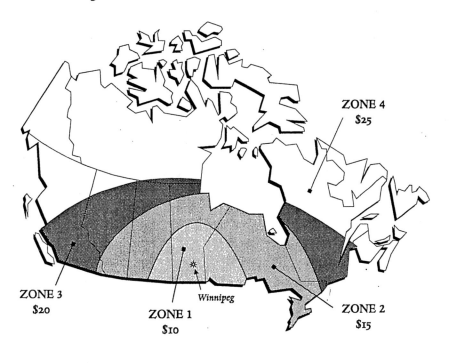

SOURCE: Dale Beckman, David Kurtz, and Louis Boone, *Foundations of Marketing*, 5th ed. (Toronto: Dryden, Holt, Rinehart and Winston of Canada, 1992), 372.

## ◊◊ How the Market Price Affects the Firm

What size mark-up should a producer take? What size mark-up should be allowed for wholesalers and retailers? What kinds of discounts, allowances, and rebates should be granted? How will freight be handled? While each of these questions require some type of policy decision, it is always important to realize that every policy decision will have an effect on the financial health of the firm. A price set too high will not be accepted by consumers. A price set too low will not allow the firm to be profitable.

In classic microeconomics, the consumer is represented by a demand curve. While these are very difficult to estimate, one does need to think about the consumer as one makes each of the policy decisions listed in the previous paragraph. One technique that takes the consumer into consideration is break-even analysis.

Calculating a break-even point is fairly easy. It is the intersection of two lines – the total-revenue line and the total-cost line. The total-revenue line has a slope equal to the market price per item and no intercept (as there is no guaranteed revenue if no units are sold). The total-cost line has a slope equal to the average variable cost per item and an intercept equal to the total fixed costs (see fig. 3).

If we have a market price, it is fairly easy to calculate the break-even point. Any units that can be sold in excess of the break-even point will be profit for the firm. Any shortfall in sales will result in losses. As the total-cost line is unaffected by the market price, the only detail that can change the break-even point is price differences. Higher prices represent more steeply sloped total-revenue lines and a lower break-even point in units to be sold. Lower prices have the opposite effect. Now with different break-even points calculated for different market prices, the marketer's understanding of the consumer comes into play. Will consumers buy enough units at the higher price or the lower price for the company to break-even?

Break-even analysis is deceptively simple. It is, in fact, one of the most powerful techniques in a marketer's arsenal of analytical weapons.

*Figure 3.  BREAK-EVEN ANALYSIS*

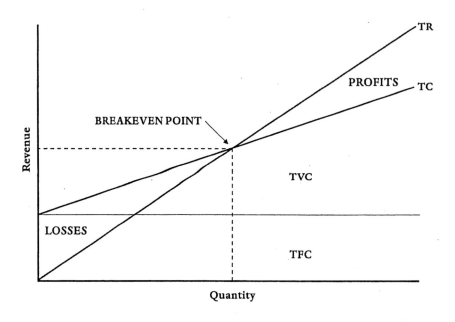

$$\text{Break-even point in units} = \frac{\text{Total Fixed Costs (TFC)}}{\text{Price (P)} - \text{Average Variable Cost (AVC)}}$$

# PROMOTION STRATEGY

If a tree falls in the wilderness and no one is present, does it make a noise? While this old philosophical question has been debated for centuries, it is easy to answer when it is adapted to a modern marketing context. If a company advertises in a magazine that nobody reads, has it communicated with anyone? The answer is no.

Promotion is all about communications. Any messages that a company sends to inform, persuade, or remind customers as they make a purchase decision perform a promotional function. This communication task can be performed on a one-to-one basis (a salesperson talking to a potential client) or on a one-to-many basis (an impersonal television advertisement broadcast to millions of viewers).

A promotional mix is ultimately a blend of personal and nonpersonal selling choices. Along with advertising, nonpersonal selling involves public relations and the field of sales promotion, some of the tools of which are coupons, free samples, contests, and in-store displays. Unlike the stereotypical vacuum-cleaner salesperson of an earlier era, the modern salesperson must understand consumers and help them to satisfy their needs.

Producers of goods or services must investigate four areas: the communications process; the choice of advertising medium; the field of sales promotion; and the personal-selling process. Let's examine each of these in more detail.

## ◊◊ The Communications Process

Communications begins with a sender, generally a company with a message it wishes to relay. The message can be either a piece of information, a question or request, or an opinion or piece of advice. Good messages must: gain the attention of a receiver; be understood by both the receiver and sender; and stimulate the needs, change the attitudes, or reinforce the learning of a receiver, and suggest an appropriate action or reaction.

For communications to be effective, there must also be a receiver. This could be a single individual or millions of people. In deciding how to reach the receiver, the firm must choose a transfer medium with which to deliver the message. This medium can be a salesperson or a television commercial or a flyer or a Facebook page or a free sample brought to someone's door. Generally, the more receivers that can be reached with a transfer medium, the lower the per-capita cost for its use. During the Super Bowl, television-commercial time may cost four million dollars but reach hundreds of millions of people for a per-capita cost of pennies per person. A flyer delivered by Canada Post will reach far fewer people and its cost, as third class mail, can be measured in tens of cents per person.

Messages must be encoded or translated into understandable terms. A company with a new brand may want to build awareness. Its message may be nothing more than, "Hello Canada. I want you to meet my new brand." The translation of that message is done once a transfer medium is chosen. If a company chooses a television commercial to promote its product, it might use a recognizable celebrity, humour, and rock-and-roll music. If it decides on personal selling, then it might use a four-colour flyer, a gift pencil, and an attention-getting demonstration. The message must be decoded by the receiver. While watching a television commercial for Obsession cologne, you may wonder what message you are supposed to extract. A recent public-service message on the dangers of steroid abuse was tested on a group of teenagers. When asked what message they had received, most identified it as "Taking steroids makes you more muscular." The message about the dangers of steroids had been either encoded or decoded incorrectly.

An enemy of good communications is noise, which can be defined as anything that interferes with the transmission of a message or reduces its effectiveness. Noise can be something as simple as the static you hear while listening to a radio station or the sound of children playing in the room where you are watching television. It can also refer to the thousands of marketing messages we receive every day. As we cannot possibly comprehend them all, we build up perceptual screens. These are an impediment to marketing communications. We also use our television remote-control to filter communications.

Marketing communication is not one-way, or simplex, from the manufacturer to the consumer. If one truly embraces the marketing concept, one knows that consumers also need to be heard. They provide much-needed feedback on a producer's message. Perhaps consumers buy a product, perhaps they don't. After buying the product, they may have problems or suggestions about how to improve the product that need to be communicated to producers. WATS lines (1-800 numbers), social media posts, and marketing research studies are just a few vehicles that allow consumers to be heard.

Of course, marketing communication is not a one-time process. It is a never ending loop of sending and receiving (see fig. 1).

## ◊◊ Choosing Between Advertising Media

Given the cost of developing an advertising campaign, the correct choice of a medium is critical. The key is to determine, through research, a specific description of the target market, including a determination of its size, key characteristics, attitudes, opinions, and preferred activities. One must

MARKETING INSIGHTS

balance the cost of an advertising medium with the reach or coverage of the target market. Overcoverage (spilling into unwanted target markets) can be as bad as undercoverage.

*Figure 1.*   THE COMMUNICATIONS PROCESS

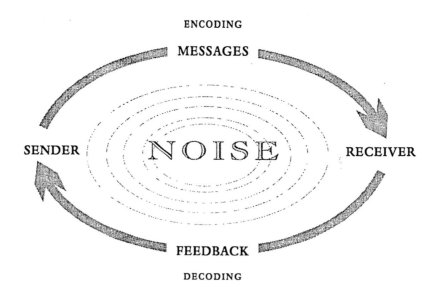

The largest fraction of advertising dollars, about 38-percent, goes to television – network and cable combined. Television's share of advertising dollars changes little annually, in part, because the quantity of its advertising is fixed. Typically, no more than eight minutes of advertising are allowed during a thirty-minute television broadcast. With the supply fixed and demand growing, it comes as no surprise that the cost of television advertising time is growing faster than the rate of inflation. Television is popular because of its impact on a range of senses. A television advertiser can combine sound, motion, and colour to create advertising with impact – impact that might break through our perceptual screens. Also, one national television advertisement reaches many millions of people and, through repetition, has the potential to reach many millions more. Of course, the costs of producing a television advertisement (sometimes as much as five million dollars) can be prohibitive for many companies. Television commercials rarely run longer than thirty seconds, and some only last fifteen seconds. A company may have to pack its commercial with information, but because that information is so dense, the viewer could have trouble retaining it. Finally, the effectiveness of television commercials decays quickly. Most have a life span of thirteen weeks; they must then be retired. Some lose their effectiveness because the public do not trust the actors or the information presented.

Internet advertising receives 30-percent of advertising dollars annually. This has been the fastest growing medium starting with virtually no spending in this area fifteen years ago. Like a magazine, some websites have banner ads at the bottom, top, or sides of a page. When you search for information, engines like Google, Yahoo, or Bing suggest some sites because they are paid to do so. Advertisers can spend a fixed amount for an advertisement to appear or they can pay-per-click only paying when a viewer clicks on its ad. The effectiveness of Internet advertising is still uncertain. Some automobile companies have discontinued banner advertising as they found many people simply clicked accidentally on their banner ads. For some free Internet services, like Facebook or LinkedIn, Internet advertising is the revenue stream which allows it to survive. In the first quarter of 2016, Facebook generated a little over $5.00 per user per quarter in advertising revenues.

About one-eighth of advertising dollars goes to newspapers. They suffer from a short lifespan (often less than a day), the fact that most readers skim them quickly (usually less than twenty minutes), and the smudgy quality of their images and text. Though inks have been improved in recent years, and full and partial colour are more easily used, newspaper images cannot compare with those of magazines and other periodicals. On the plus side, newspapers reach nearly 90-percent of households; readers can refer back to previously read sections, or can keep advertising handy until it is needed (say for shopping); and newspaper advertising can be easily moved from one day to another, or from one page or section to another, or from one city to another.

Magazines and periodicals are the fourth most popular medium, garnering 7.5-percent of advertising expenditures. Because magazines are designed for very specific target markets, advertisers are given an excellent range of consumers to target. Just imagine the diversity of readers for *Architectural Digest*, *National Geographic*, *Seventeen*, *Soldier of Fortune*, *GQ*, and *Out*. Magazines have longer life spans (typically thirty days), and a better quality of image reproduction than newspapers. Magazines lack the flexibility of placing advertising in certain geographic markets and not in others. Magazines need a certain minimum subscription base to remain in business. A magazine designed for the north end of Saskatoon just wouldn't have enough readers to justify its existence.

Radio has seen its share of advertising dollars decline since its heyday in the 1930's and 1940's. With the advent of television, radio became a medium of music and talk shows. Today radio attracts only 7-percent of advertising dollars spent. Like television, radio suffers from an overabundance of stations, and this in turn, fragments the market. In Windsor, one can listen to more than thirty radio stations, both FM and AM, originating in both Canada and the United States. If an advertiser wants to reach a specific type of listener, it might have to advertise on several different stations. People cannot retrieve a radio ad (or a television ad) once it has been played. Information such as telephone numbers and addresses are hard to convey. Radio advertising can, however, be created quickly and changed within a few minutes. Radio advertising time is very cheap to purchase (a few hundred dollars for a thirty second spot on a top-rated station), and radio is a portable medium. People listen to a radio in cars, on boats, at the beach, in a gym, and so on. They can't do the same with television.

Direct mail involves a broad spectrum of advertising media, including flyers, catalogues, postcards, sales letters, folders, and booklets. It is difficult to measure the percentage of advertising expenditures for direct marketing as companies that exclusively used paper mail have converted to electronic mail. While many people call these solicitations junk mail, it may surprise you to know that 3-percent of people receiving a piece of direct-mail advertising decide to make purchases based on the information that this advertising provides. Direct mail allows an advertiser access to every home in a given area (it thus reaches even more people than newspapers do), thanks to Canada Post. It can provide a large amount of detailed information to a customer, yet the format of an ad can be changed easily and quickly should an advertiser so desire. At a slightly increased cost, target mailing lists (made up of people who have a common attribute) may be purchased allowing for a personalized message. Even though most direct mail is third class, the per-capita cost of postage makes direct mail the most expensive of all the media. The quality of a mailing list is also critical. If a database of addresses is correct today, six months from now 10-percent of the addresses will have changed. Mailing lists which are out-of-date lead to much undelivered mail. Maintaining an accurate database of addresses is both a difficult and expensive proposition.

Finally, there is outdoor advertising. Posters, billboards, electronic displays, signs on the sides of buses and taxis, and ads on transit shelters account for about 6.5-percent of advertising expenditures. If designed properly, outdoor advertising can effectively communicate short, simple messages. In high-traffic areas, it is possible to expose consumers to an advertising message repeatedly. Outdoor advertising is especially effective for notifying potential customers about nearby sales (just witness the proliferation of outdoor advertising for summer garage sales). Of course, the message must be brief. It is impossible to convey an advertising message, an address, and a phone number on a billboard. Environmentalists have begun to question the use of outdoor advertising, identifying it as a form of visual and spatial pollution.

It should be noted that, regardless of the medium, the advertising message changes over the product life cycle (see fig. 2).

*Figure 2.* THE LINK BETWEEN THE PRODUCT LIFE CYCLE AND THE ADVERTISING MESSAGE

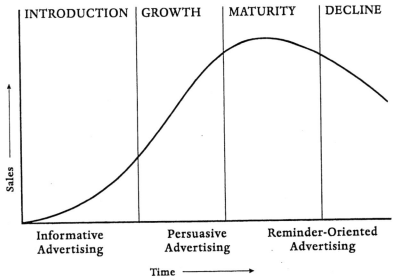

SOURCE: Dale Beckman, David Kurtz, and Louis Boone, *Foundations of Marketing*, 5th ed. (Toronto: Dryden, Holt, Rinehart and Winston of Canada, 1992), 509.

## ◊◊ The Field of Sales Promotion

When asked to define promotion, most people immediately begin talking about advertising. But the importance of advertising in the promotional mix has been declining of late. No doubt this is a direct result of the proliferation of new cable-television stations, greater reach of radio stations, and the growth in social media on the Internet. But advertising's loss has been sales promotion's gain. It is difficult to define exactly what is meant by a sales promotion. Some define it by what it is not. Sales promotion is not advertising, public relations, or personal selling. Others see sales promotion as an attempt to reach a very specific target market and motivate its members in very specific ways.

One of the most important sales-promotion techniques involves the distribution of samples, coupons, and premiums. Samples are products that are distributed for free. They are intended to build awareness, demonstrate a product's features, and eventually stimulate trial purchases. Distribution may be done through the mail or on a door-to-door basis. You may have recently received a sample of a new shampoo or a new laundry detergent.

Coupons offer a specific price reduction on the next purchase of a product. They are especially useful in stimulating trial purchases, and can help to overcome the perceived price barrier that prevents some consumers from buying certain premium products. Most retailers readily redeem coupons – which are distributed through the mail, in magazines, inside packages, and in newspapers – as they are not only reimbursed for the discounts but they also receive a handling fee.

Premiums are bonus items given free with the purchase of another product. As children, many of us begged our parents to buy a special brand of cereal because of the prize that could be found inside. In some instances, these premiums have become desirable in and of themselves. Witness the growth of sports cards, which were once used to sell bubblegum. Today the gum has all but disappeared from the packages. Premiums are used to stimulate trial purchases of a product and to initiate brand switching.

Another sales-promotion technique is point-of-purchase advertising. While many consumers enter a store with some idea of which brands they will purchase, it is often possible to influence them at the place where the purchase decision is made. Point-of-purchase advertising can be as simple as a small sign or sticker attached to a shelf or as complicated as a stand-alone display designed to show the product in a special light. Both types of promotion are designed to draw attention away from competitors' products.

Contests are a form of sales promotion that is usually intended to encourage repeat purchases. As the odds of winning a prize are fixed, the more times one enters a contest the more likely one is to win. Often contests offer significant cash or merchandise prizes to call attention to particular products. Just think of the millions that can be won in the Publisher's Clearinghouse contests. Contests may be regulated at the municipal, provincial, and federal government levels, so experts are often hired to guide a company planning this type of promotion. Generally, governments insist that the purchase of a product not be required to enter a contest, though this fact is not generally known to most consumers.

A subtle form of sales promotion is specialty advertising, which uses everyday articles to carry the name, address, and advertising message for a company. Look at the pen or pencil on your desk, the book of matches by the fireplace, the calendar on the wall, your key ring, a shopping bag, or the T-shirt you are wearing and you may find some form of specialty advertising. It turns common items into advertising media.

Finally, there are conventions, large gatherings where wholesalers, retailers, and, sometimes, the public have a chance to meet, discuss trends in a certain industry, and view displays and product demonstrations. You may have attended a car or boat show to marvel at some of the new technology. Other shows, such as those for the landscapers, retail fashion buyers, and personal-fitness-equipment-makers, are not accessible to the general public. While no significant cash-and-carry sales are made at such shows, wholesale and retail orders are often placed for the selling year ahead.

## ◊◊ The Personal Selling Process

While some are more gifted when it comes to meeting clients and pursuing sales, all people are salespeople at various points in their lives. We often try to sell ourselves, whether it is to get a job, get elected to an office, attract a mate, or seek a bank loan: selling is not an easy task. It involves seven distinct steps.

The salesperson begins the process by identifying potential customers. Prospecting for customers, he or she may have to consult databases, talk to friends, or networking through social clubs. Generating these lists is a difficult job, and the salesperson does not often receive an immediate payback for it. Identifying potential customers is a continual part of the life of a salesperson, and it can be frustrating. In the qualifying phase, salespeople must check their potential-customer lists to ensure that they are not wasting valuable time talking to someone who is not a customer and is unlikely to become one. He or she needs potential customers with not only a desire to purchase but also purchasing power and the authority to buy.

Next comes the challenging step of actually meeting with a potential customer. Making the approach is never easy, and the difficulty increases with the importance of the customer. In other words, a customer who might place an order for three million dollars of equipment is much harder to approach than a customer who might place an order for a few hundred dollars. To ease the tension, a salesperson is well-advised to learn as much as possible about these prospective customers and the environments in which they operate.

When the formalities and pleasantries of a first meeting are out of the way, a salesperson needs to give the sales message to the potential client. In making the presentation, the salesperson describes the product's major features, points out its strengths and benefits, and lists previous successes. Above all else, the presentation should be clear, concise and positive.

Because some product benefits are difficult to describe, it is sometimes necessary to supplement the presentation with a demonstration. The salesperson can also reinforce and supplement what he or she has said by showing the product in action. It is important for the demonstration to be unique. Anyone can throw some dirt on the floor and then vacuum it up. It is much more effective to show that the vacuum can remove spills of both liquid and dry material. However, many sales have been lost when a product malfunctions or the salesperson cannot demonstrate all the features. Clearly, careful planning and practice are the keys to successful demonstrations.

No salesperson can fully anticipate all questions a potential customer will ask.  The presentation or demonstration may not lead to an immediate sale.  The potential customer may have some questions about the product or some objections to making the purchase.  Answering these questions and handling these objections should not be seen as a chance to remove barriers, but as the opportunity to present additional information.  If someone says, "I don't like green," they are actually saying, "In what other colours is the product available?"

Every presentation inevitably reaches a point when the salesperson must ask the potential client to make a purchase.  Closing techniques vary considerably from one salesperson to another.  Some prefer to present potential clients with two equally acceptable (to the salesperson) alternatives, so that either outcome is good.  Others prefer the higher risk strategy of silence, which will always force a customer decision – though not always the one a salesperson desires.  Still others prefer to make appeals to emotion or guilt to swing a customer and make a sale.

But the sale is not the end point of the selling process.  From psychology, there is a concept called cognitive dissonance.  After making a purchase, customers may experience post-purchase doubts about the product or the size of the order.  Left to themselves, these customers may well cancel the order or reduce its size.  A good salesperson visits the customer after the closing, and makes certain that all is going well.  The goal of a salesperson should not be to make one sale but, rather, to develop an ongoing relationship that leads to many sales.  If the salesperson psychologically reinforces the customer's decision and handles any problems with delivery or order specifications, she or he will ensure customer satisfaction, and will increase the likelihood of further purchases.

Both personal and non-personal communications play important roles in the promotion mix.  Studies have shown that leading up to and following a transaction, non-personal communications plays a relatively more important role.  When the transaction is actually taking place, personal communications is much more important (see fig. 3).

# DISTRIBUTION STRATEGY

When engaging in polite dinner conversation, it is said, one should avoid three topics: politics, sex, and religion. To those add Canada Post. Few Canadian institutions can evoke such strong emotions at the mere mention of their name. Much of the ill will generated by Canada Post can be traced to its distribution strategy.

The cornerstone of any distribution strategy, or any marketing strategy, is the consumer. A company should set a standard of distribution upon which consumers can rely. Such standards normally include a time frame and a penalty clause, should the standard be breached. Some pizza-delivery companies have a standard that promises customers a free pizza if they are kept waiting longer than thirty minutes. A major road-equipment company guarantees spare-parts delivery within forty-eight hours or it pays the repair bill.

Canada Post, too, has a customer-service standard for correctly addressed mail. Within the same city, a letter should be delivered within two working days. Between cities in the same province, a letter should be delivered within three working days. Between cities in different provinces, a letter should be delivered within four working days. Do you notice anything missing from this standard?

There is no penalty clause. In 2007, Canada Post complied with its customer-service standard 85-percent of the time. By 2013, this had improved to 99-percent of the time. This seems an impressive change, until you realize that Canada Post handles nearly 9.0 billion pieces of mail each year – 4.0 billion pieces of letter mail and 5.0 billion parcels. Ten-percent of the mail originates with individuals; the remainder with businesses. Still, the 1-percent of mail delivered late is equivalent to ninety million pieces. Every year, one in two Canadians will have a piece of mail delivered to them late.

It is this uncertainty about prompt delivery that makes Canadians unhappy with their postal service. If delivery were slowed, few people would be upset as long as they could be guaranteed that the mail would arrive within the specified time frame. Canada Post's competition, the courier companies, back their guarantees with self-imposed penalties if they fail to meet service standards. To win us back, Canada Post may want to consider a similar gesture.

## ◊◊ Physical Distribution

We begin our discussion of distribution by focusing on the activities that move products from manufacturers to the ultimate consumer. After all, physical-distribution costs represent half of all marketing costs incurred by a firm. Logistics, or physical distribution, involves transportation companies, warehousing firms, financial institutions, insurance companies and marketing-research firms. It also involves customer service, inventory control, materials handling, order processing, transportation, packaging and warehousing (see fig. 1).

The starting point in logistics is defining a customer-service standard.

Too low a standard may mean dissatisfied customers and lost sales, while too high a standard may mean increased job stress and high distribution costs. A proper standard requires a trade-off between the customer and the manufacturer. Tied to the customer-service standard is the choice of transportation mode: train, truck, boat, airplane, or pipeline. Though these will be discussed as separate alternatives, companies must often combine them to meet customer needs. For example, shipping product to Prince Edward Island from an Ontario plant might require a truck, train, and boat.

Figure 1.   COMPONENTS OF A PHYSICAL-DISTRIBUTION SYSTEM

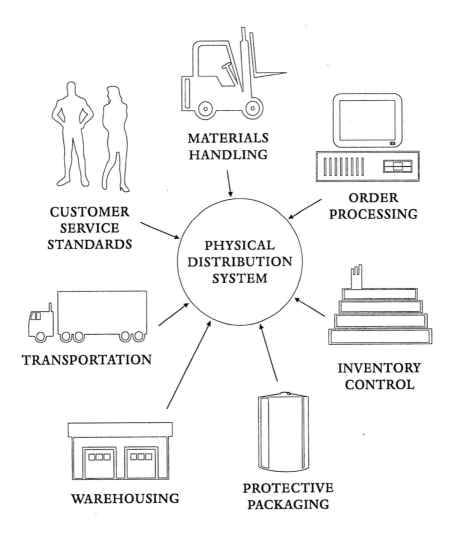

Trains move the most product and have always been most efficient at moving large quantities of bulk goods over great distances. Their role has diminished over the last several decades due to two factors. First, every manufacturing plant used to be built near a railway so that a spur line could service the plant. Today, most industrial parks are built away from railways and close to roads. Second, the size of loads being shipped has declined. With just-in-time inventory systems, smaller quantities of products are being ordered more frequently. Half–filled railway cars are not efficient to move, and are often the cause of damaged goods and shifted loads. The fact that the railway has retained its lead can be attributed to its flexibility in moving liquids, raw materials, gases, grains, and finished goods very safely.

The trucking industry has grown at the expense of the railways. Due to the host of sizes (from panel vans to eighteen wheelers) and types (for example, refrigerated vehicles or tankers) available, there is great flexibility in the size and type of loads trucks can carry. They can reach any company that is on a road. They are excellent for quickly and frequently moving products short distances. Perhaps trucking's greatest failure is its human component. Drivers cannot drive twenty-four hours a day; they need to sleep and eat. Sleeper cabs allow for teams of drivers, and therefore improve productivity, but they are not a perfect solution. Trucks also need to be refuelled and maintained. Tires can go flat and, no matter what automobile drivers may think, trucks are constrained by speed limits. If a shipment of lobster must travel from New Brunswick to Alberta in one day, no truck can get it there on time.

Boats are best for moving large quantities of product over long distances. Most international trade requires the use of oceangoing vessels, whether it be to bring oil from the Middle East or cars from the Far East, or to take grain from western Canada to the rest of the world. Canada is blessed with excellent inland waterways. Smaller vessels move products through our Great Lakes and up our rivers. Boats tend to be the slowest mode of transportation, and thus are also the cheapest. Airplanes, of course, are the opposite. They offer the fastest way to move products and the most expensive. Planes are used when speed is crucial. They transport perishable products such as flowers and seafood; products needed in emergency situations such as spare parts for machines and human organs for transplants; expensive items such as gemstones and designer fashions;

and for important documents and parcels.

When one thinks of pipelines, oil immediately comes to mind. But pipelines can be used to move other liquids, gases and even some solids. Pipelines are not a common mode of distribution, yet much research is being conducted to see if pipelines can be more efficiently exploited. Pipelines generally provide a one-way movement of goods – from the source to the receiver. It is already standard operating procedure to place some items (for example, documents or test samples) in capsules that are then carried along the pipeline to the receiver. There have been experiments to determine whether two types of materials can be transported in the same pipeline (oil and water, for instance). Some people foresee a day when a reverse flow will be possible in a pipeline, thus allowing the two-way transportation of goods.

## ◊◊ Inventory Control

Manufacturers, wholesalers, retailers, and consumers are concerned with the question of how much inventory to have on hand. Again, this is a cost-trade-off question. Every time an order is placed, there is a cost. On one hand, assuming that the cost is fixed, that it remains unaffected by the size of the order, it makes sense for companies to place large orders. On the other hand, inventories are generally purchased with borrowed money. If its inventory can be sold quickly, a company's capital will be freed. It will be able to use it to repay the loan, and interest charges will be minimized. If the inventory moves slowly, interest payments grow, a company's credit rating falls, and the cost of borrowing rises. This suggests that perhaps it would be best to place many, smaller orders.

If the two costs are added, a parabolic total-cost curve is found (see fig. 2). As you may remember from your calculus classes, a parabola has a minimum point. After finding that minimum point, a company determines the economic order quantity – the amount that should be ordered at any one time to minimize inventory costs. If it knows what the approximate yearly demand for a product is, the company can also calculate how often it should place orders. Suppose the economic order quantity is five hundred cases per order, and the company knows it needs six thousand cases per year. Twelve orders would be placed annually, or approximately one per month.

*Figure 2.  MINIMIZING INVENTORY COSTS*

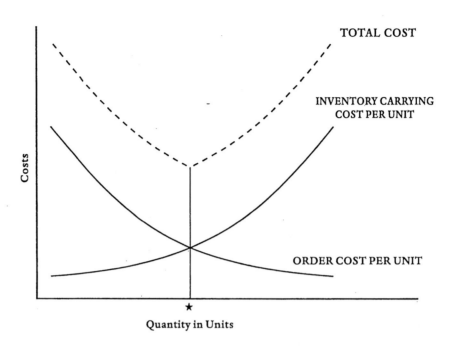

EOQ (economic order quanitity) formula, to calculate the minimum point of the parabolic total cost curve:

$$\sqrt{\frac{2RS}{IC}}$$

Where:   R = the annual rate of usage
         S = the cost of placing an order
         I = the annual inventory carrying cost percentage
         C = the cost per unit

SOURCE: Dale Beckman, David Kurtz, and Louis Boone, *Foundations of Marketing*, 5th ed. (Toronto: Dryden, Holt, Rinehart and Winston of Canada, 1992), 397.

## ◊◊ Materials Handling

In the 1940's, every case of every product was moved by hand. Imagine a truck reaching a dock and being unloaded by hand. Each case is passed from person to person and then reloaded, again by hand, onto a boat. Picture the weather as cold and wet. Finally, imagine the amount of damage, spoilage, and pilferage that could occur when products are handled in this way and under such circumstances.

Today, partly due to the costs of damage, pilferage, and spoilage and partly due to the cost of labour, human handling of products is minimized. Conveyor belts have been designed to move products across warehouses.

Stackers have been designed to place products on pallets (one-metre-square raised wooden platforms) and forklift trucks move the pallets onto trucks, railway cars, and boats.

Another important innovation has been the container. Shifting product from one mode of transportation to another (for example, from truck to boat) increases the risk of damage, pilferage, and spoilage, even if equipment is used. Special containers have been designed that can sit on the flatbed of a railway car, on the trailer of a truck, or in the hold of a ship. Once loaded, containers never need to be unloaded until they reach their destination.

## ◇◇ Warehouses

There are two ways to think about warehouses. One is to focus on the level of activity they maintain. In September and October, tomatoes are harvested in Ontario and taken to local canneries. At that time, all the tomato juice, tomato ketchup, tomato sauce, and tomato soup for the next twelve months is manufactured. If you buy a can of juice or a bottle of ketchup in July, it is probably several months old. This is not a very appetizing idea. Where has it been? A storage warehouse is designed to be a relatively inactive place where products are kept for relatively long periods of time – months. On a day-to-day basis, few trucks arrive and few orders are filled. A distribution warehouse, by contrast, is a very active place. Small quantities of all the products in a company's product line are stored there for only a few days. During this time, the products are sorted and prepared for shipping to fill orders received from wholesalers or retailers.

A second way to think about warehouses is as efficiency centres. Suppose a company in Montreal had orders to fill in Regina, Saskatoon, Edmonton, Calgary, and Winnipeg. It could load five different trucks, one for each city, and wait for them to make their deliveries. But that would not be very efficient. For most of the journey, the five trucks would cover the same route, only diverging towards the end. It would be more efficient to use a larger carrier, say a railway car, to ship all the product to a central site in the West. The warehouse at the central site would then break the order into the five smaller lots to be delivered by truck. This activity would occur at a **break-bulk** warehouse.

At a **make-bulk** warehouse, small shipments from several different producers destined for the same city are assembled for efficient transport by a larger carrier. If producers in Waterloo, Hamilton, Guelph, and Mississauga all have shipments bound for Halifax, it would be cheaper for them to put those shipments together on a large truck rather than using four smaller trucks.

## ◊◊ Marketing Intermediaries

Wholesalers and retailers operate between the producer and the consumer. Although they lengthen the distribution channel and slow the distribution rate, these intermediaries perform several essential functions.

They provide information to manufacturers about the reactions of retail stores and consumers to new products, about customer complaints, and about competitor activities. On behalf of a manufacturer, they can be a source of technical information, as well as information about price changes and industry trends to retailers and consumers. Intermediaries often provide financing for their customers. To encourage you to buy, **Sears**, **WalMart**, and **The Bay** offer you the convenience of a store credit card. A wholesaler often extends credit to a retailer. Should the retailer be able to sell its inventory, it can repay the wholesaler and claim the rest toward overhead and profit.

Wholesalers and retailers perform a storage function. Every retail store maintains an inventory of products both on its shelves and in its storeroom. Likewise, wholesalers need to keep a supply of product on hand to fill orders from retailers. These distributed inventories allow manufacturers to minimize their inventory costs. Intermediaries also perform a transportation function. Sears will deliver a new refrigerator to your home. Wholesalers will deliver orders to the door of retailers.

It is not easy for a manufacturer to make the decision to lengthen the distribution channel by adding intermediaries (see fig. 3). The decision can be influenced by the type of product being sold. Perishable products need short distribution channels. So do high-fashion products. Standardized products or low-unit-value products typically have longer channels. The type of market influences the length of the channel. Consumers like to purchase products from retail stores. Industrial customers often prefer to deal directly with manufacturers. In densely populated areas, a bakery can sell directly to consumers, but in low population-density areas, wholesalers and retailers are used.

*Figure 3.* FACTORS AFFECTING THE LENGTH OF A DISTRIBUTION CHANNEL

❖ The Consumer/Customer
❖ Product Characteristics
❖ Manufacturer Characteristics
❖ Factors in the Business Environment

Certain characteristics of the manufacturer can have an impact on the channel. Companies that want the control which comes from having their own wholesale and retail outlets must have large financial resources. Most small or new manufacturing companies have to rely on the kindness of intermediaries. A manufacturer may also desire total control over the presentation of its product. In that case, a shorter channel is used. Finally, the manufacturer must consider the types of channels allowed by law and those used by its competitors. **Molson** and **Labatt** would like to sell beer through corner variety stores in Ontario, but they must use the intermediaries provided to them by law. Mary Kay, of **Mary Kay Cosmetics** fame, claims that when she wanted to enter the cosmetics industry, **Max Factor**, **Maybelline**, and **Cover Girl** had a virtual lock on retail stores. She chose a more direct route, party selling, to give herself a competitive edge.

◊◊ Wholesaling

Wholesalers are companies that sell to retailers, other wholesalers, and industrial customers, but do not sell significant quantities to individual consumers. A manufacturer may wish to forward integrate or a group of retailers may wish to backward integrate by acquiring a wholesaler, but most wholesalers are independently owned and operated.

Independent wholesalers can be divided into two groups: those that take title to, or own, the products they handle and those that do not. The latter group is made up of agents and brokers. One example of this group is an auction house. In London, England, one of the world's famous auction houses, **Christie's**, displays and sells precious art. While the art is stored at the auction house in the days leading up to the auction, it remains, at all times, the property of someone else.  As an agent-broker, **Christie's** must bring buyers and sellers together, and for this it receives a commission (as much as 20-percent of the selling price). When agents and brokers are mentioned, many think of stockbrokers or real-estate agents. Agents and brokers may only work on one side of a transaction. I want my stockbroker to get the highest possible price for my stock.  You want your broker to get the lowest possible price for the stock. If the broker worked for both of us, he or she would have an insurmountable conflict of interest.

Merchant wholesalers take title to the goods they sell to others.  For example, a truck wholesaler takes title and possession of perishable products like bread, milk, vegetables, fruit, potato chips, candy, and eggs. This wholesaler arranges frequent, perhaps daily, deliveries to retailers to guarantee the freshness of the goods being sold.  If the truck wholesaler miscalculates and buys more eggplant than is demanded by retailers, it must find a way to dispose of the excess at its own expense.

## ◊◊ Retailers

Retailers sell products and services to the ultimate consumer for his or her own consumption. Many different classification systems have been developed in an attempt to better understand the host of different retailers. An easy way to classify a retailer is to consider the amount of consumer shopping effort it demands. Convenience stores are designed for maximum accessibility. Long store hours, fast checkout service, ample parking, and easy-to-reach locations are hallmarks of the variety store, gasoline station, dry cleaner, grocery store and instant banking machines, all of which may be considered convenience retailers.

Shopping stores are designed to allow consumers to compare prices, brands, and product components before making a purchase. Typically, shopping retailers – for example, furniture stores, appliance stores, clothing outlets, and sporting goods stores – try to differentiate themselves through floor layouts, window displays, special merchandise or brands, and knowledgeable salespeople.

Consumers must work to shop at specialty stores. Often they have shorter hours of operation, no free parking, and less accessible locations. But consumers are willing to expend the effort because of a particular specialty store's unique combination of product lines, service, and reputation. Specialty stores often work to breed loyalty among consumers, so that repeat business is guaranteed.

There are some special decisions retailers must make (see fig. 4). The first concerns a store's image. The moment customers walk into a store, they form an impression about the type of shopping experience they will have there. Imagine yourself entering a men's clothing store. The staff offer you a cup of cappuccino or, perhaps, some sparkling water. You are invited to discuss your clothing needs while seated on a leather couch. The store's walls are covered in walnut panelling with brass accents. This is shopping at its best. Compare this to the worn linoleum tiled floors, chrome racks, and dark ceilings that you might see at a **Giant Tiger** discount department store. Image is a very powerful promotional tool for a retailer.

Figure 4. WHY CONSUMERS CHOOSE A PARTICULAR RETAILER

- Low Prices
- Convenience of Location or Hours
- Variety of Selection
- Perceived Product Quality
- Assistance from Salespeople
- Reputation for Integrity and Fairness
- Special Services Offered
- Sales/Special Value Offered

Equally important is a store's location. While shopping malls, which offer one-stop shopping, have grown in popularity, not every retailer is wise to locate in one. If a store's products appeal to impulse buyers, then a location that gets lots of walk-by traffic is the most desirable. What are traffic patterns like? Is there a bus stop nearby? What are the neighbouring stores? It wouldn't be wise to open a **Ralph Lauren Polo** shop beside a retailer called **Everything for a Buck**.

## ◊◊ Intensity of Distribution

If a manufacturer has chosen to include wholesalers and/or retailers in the distribution channel, it must decide how many of each it needs? This raises the question of market coverage or intensity of distribution. Exclusive distribution comes into play when the rights to sell a product are limited to a geographic region. In Winnipeg, there may be only one store that carries **Waterford** crystal, or **Royal Daulton** china, or **Mercedes-Benz** automobiles. In some cities, perhaps, no one carries a specific brand of product. Exclusive distribution is often used to enhance the image of a product and build its prestige.

Intensive distribution is designed to saturate the marketplace with a product. Everywhere consumers turn, they find the product. Windshield-washing fluid, for instance, can be acquired at a hardware store, an automotive-parts store, a grocery store, a gas station, a variety store, a department store, a feed-grains store, and a discount store. Clearly, intensive distribution is used to make a product easy to purchase, and is used most often for convenience goods.

There is also selective distribution. In this case, to reduce the costs of intensive distribution yet avoid making a product hard to find – as it can be when exclusive distribution is used – a small number of intermediaries are employed. If you wish to purchase a Sony television, there are probably a handful of retail locations in your community where you can find one. When you visit these different retailers, you can compare not only prices but also the availability of credit, service terms, and delivery charges. The selective-distribution strategy is most frequently used for shopping products.

# THE CASE METHOD

Socrates, born in Athens, Greece, in 470 B.C., has long been considered one of the world's great philosophers. He founded a school of philosophy, which, in later years, influenced Plato and Aristotle. It required people to know themselves first. Socrates believed that goodness was based on wisdom, while wickedness was based on ignorance. No wise person would deliberately choose evil in the long run, but most people, through ignorance, might choose evil if it appeared to be good at a given time.

Why is Socrates of ancient Greece important to the teaching of marketing today? He employed a unique style to educate the young people of Greece. He didn't write books. We can find no trace of any writing by Socrates, and some scholars believe he wrote nothing. The only knowledge we have of Socrates comes from the writings of Plato. He preferred to wander the streets, marketplaces, and gymnasia. When he had gathered a few students, he would sit on a bench with the young people gathered at his feet. Socrates taught by questioning his students about their opinions and then asking further questions about their answers. By this means, he could show them how inadequate their opinions were. Then he helped them go beyond opinion and search out essential truths.

His approach to learning was considered dangerous. By asking questions of his students, he forced them to search for new meanings and to question the standard ways in which government and commerce were conducted. Inevitably, those in power brought charges against Socrates that he was corrupting the youth of Greece. He died in 399 B.C. after drinking a cup of poison hemlock before the state could execute him.

## ◊◊ The Case Method

This book is likely being used to supplement a textbook in an introductory marketing course. The goal of such a course is to give you a whirlwind tour of marketing – to show you dozens of concepts, get you comfortable with the jargon, and stimulate you to dig deeper. In most introductory courses, instructors will assume that students have no previous knowledge of the subject area. These instructors have a great quantity of information to share and little time in which to share it. Most instructors will use the tried and true lecture format. They will use visual aids and a chalkboard, or whiteboard. They will read from prepared notes. Having listened, you will take notes and, occasionally, when things are not clear, you will ask questions. While lecturing is an established form of teaching, there is some doubt about its overall effectiveness. Students listen for possible exam questions. If the instructor is not going to test them on the material, then they are not likely to try and master it. Even such mastery is a fleeting goal, for as soon as the exam has been completed, some students will forget the lecture material.

If mastery of marketing concepts is the ultimate learning goal, then learning by doing or learning on the job is the best approach. As you may have discovered from personal experience, it is one thing to talk about swimming, for example, and quite another to actually do it. In the last three decades, universities and colleges have instituted co-operative work-study programs and summer internships to allow students to learn on the job and apply their knowledge between academic terms. Equally popular are group projects where three to five students work with a client firm to study a problem and make recommendations. While these approaches work well for more advanced courses, for introductory courses enrolling many students, learning by doing is very difficult to coordinate.

The Socratic method allows us the best of both worlds. Using a case study, an instructor captures a real-world problem and brings it into the classroom. By assuming the role of decision maker, the student is forced to apply the knowledge being shared through the instructor's lectures. Through the questions posed by the instructor and the answers of classmates, students not only develop mastery of the material but also can begin to see the limits and artificialities of the theoretical concepts discussed. The Socratic method also builds individual communication skills, as students must logically argue and support, either in writing or orally, their positions. As students quickly discover, making decisions in the face of ill-structured problems, incomplete information, and the need to make assumptions is a messy, yet interesting, task.

The case-study method, or Socratic method, is the main feature of this book. Here, you will find contemporary case studies written about real companies from across Canada. Every case places you in the role of central decision maker. Some companies you will recognize by name. Others have been disguised to give the company anonymity. Some of the problems confront marketers every day, while others are unique. Some cases will require you to use structured, quantitative analytical techniques. Others will demand that you be more creative and qualitative in your analysis. Try to find some time to read additional case studies not assigned by your professor so that you can gain additional marketing insight.

Because these case studies present ill-structured problems, there are few set rules for their solution. In science, you are often given a series of equations that you master by solving problems. These problems are approached in the same manner. In marketing, it is not possible to find a template and apply it over and over again. With the case method, one of the frustrations you will encounter is seeing the instructor or a fellow student use a problem-solving tool which you had completely overlooked.

Worse still is watching the class discussion evolve towards a solution that is the exact opposite of the one you discovered. Don't worry about this. When acquiring any new skill, you get better and better with practice.

## ◊◊ Understanding Case Studies

Case studies are constructed along three dimensions: conceptual, analytical and presentation (see fig. 1). The conceptual dimension is the number and type of concepts to which you are exposed in a case study. At the low end of this dimension, you are exposed to one simple concept. At an intermediate level, you might be exposed to multiple concepts or a more complicated concept. At the high end, you see many difficult concepts.

The analytical dimension looks at the work expected from the student. Simple analysis clearly shows a problem and decision. Playing armchair quarterback with the help of hindsight, you decide if there was a correct match and then offer support for your opinion. At the intermediate level, the problem is presented but you must make a decision based on whatever criteria you see fit. The most difficult type of analysis only offers a situation. You must decide on the problem and a methodology for its solution.

The presentational dimension is the writing style of the case. A simple presentation style means the case is short, well-organized, and contains only the information that is most relevant to the situation at hand. At the intermediate level, bits of information may be omitted, and you will therefore be forced to make assumptions. Case writers may also introduce red herrings – misleading information. At the most difficult level, the case becomes quite long and wordy. It will be poorly- organized, much information will be missing, and what remains may be misleading.

As you can see from the case-difficulty cube, one can rate a case on each of these dimensions using a three-point scale. The simplest cases would receive a 1.1.1 rating, while the most difficult would receive a 3.3.3 rating. By adding the scores on the three dimensions, one will get a measure of overall case difficulty. The cases in this book are rated between four and seven on the overall case-difficulty scale; they are relatively short and straightforward, though here and there you will find exceptions to this rule.

## THE CASE METHOD

Figure 1.  THE CASE-DIFFICULTY CUBE

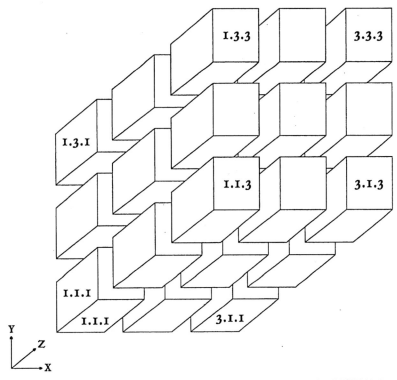

X = CONCEPTUAL,  Y = ANALYTICAL,  Z = PRESENTATIONAL

SOURCE: Michiel R. Leenders and James A. Erskine, *Case Research: The Case Writing Process* (London: University of Western Ontario, Research and Publications Division, 1989), 119.

Before beginning to analyse a case, you should remember that case studies are written to help you understand a concept or concepts. After your first reading of a case, it is worthwhile to take a minute and reflect upon the purpose of the case. What did the writer want you to learn from this case? Why did your instructor assign *this* case? If you try to get that overall sense of purpose, you can analyse the case more quickly and keep your organization logical. Occasionally, you may miss the point of a case and take your analysis in a completely different direction. Don't be concerned about this. Your ability to see problems will improve as you proceed.

◊◊  A Framework for Analysing Cases

No two instructors follow exactly the same framework for case analysis.

# MARKETING INSIGHTS

The one presented here is only a suggestion. It is to your advantage to ask the instructor about his or her approach. If possible, try to get a written outline of the instructor's expectations. Your analysis might also differ depending on whether the case is being discussed in class or submitted as a written assignment. Remember, the key to successful case analysis is maintaining a flexible approach.

Clearly, the first step is to read and re-read the case. During the first reading, try to get a feel for the situation and an understanding of the basic facts of the case. Keep an open mind as you are reading and avoid reaching too firm a conclusion about what the company should do next. You will want to differentiate between facts and opinions. In every situation, there is a certain amount of insider information which may or may not be correct. Every assertion should be supported with facts or data; otherwise it is suspect. People inside a company do not always see the problems they face in an accurate way.

During the second reading, you might wish to use several markers to highlight specific types of information – one colour for problems, another for environmental issues, another for marketing-mix variables, and so on. Carefully examine each exhibit of the case. Generally, the exhibits have been included with a purpose in mind. They contain some information that will be useful to you in analysing the situation. In some cases, the raw information may need to be processed using a special analytical technique – for example, product life cycle, breakeven analysis, and the BCG matrix. You should begin to think of various techniques and whether they are appropriate in this situation. With this initial reading completed, set aside the case and work on some other non-marketing problems. Your mind will continue to process facts and questions in your subconscious.

You are now ready to apply your instructor's case solution model. One approach is presented in figure 2. There are five steps required in the analysis. The first is to identify the problems or objectives facing the decision maker. In identifying a problem, you should be looking for root causes rather than symptoms. If the problem is a decline in sales, the root causes could be lack of product awareness, consumers having bad experiences with the product, competitor activities, problems with distribution or changes in consumer behaviour. Students complain that companies don't always have problems. A case study involving a problem-free company might focus on a new corporate objective. Perhaps the company would like to expand to new markets or develop new products. Some case problems/objectives will be simple, others will be complex. Though this first step constitutes only a small part of the analysis, do not rush it. To quote an old cliché, a well-defined problem is half-solved.

*Figure 2.*   CASE-SOLUTION MODEL

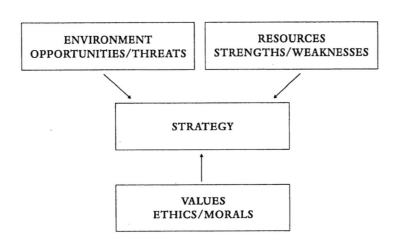

The second step requires identification of the *current* marketing strategy. More specifically, you should identify the current company policies involving each of the four P's of marketing and the target market for the product. Before you try to solve problems and change strategies, it is important that you determine the current strategy – you cannot plan a journey if you do not know the starting point.

The third step is to identify the strengths and weaknesses of the company and the opportunities and threats in the environment. Before one makes any attempt to change the marketing strategy of a company, one needs to understand the capabilities of the company and the peaks and valleys in the playing field. Students often have trouble separating environmental issues from company issues, but this separation can, in fact, be made quite easily: an issue falls into the environment if the company cannot control it. Companies cannot control governments, suppliers, consumers, and competitors: they may be able to influence them, but they can't control them. Some students also have trouble with the word *opportunities*. Students frequently will tell me that a given company has the opportunity, for instance, to expand into Mexico. That is not an opportunity or threat in the environment. What the company has, most likely, is an alternative to pursue. When considering the strengths and weaknesses of a company, look at the functional areas of finance, information systems, production, human resources, and labour relations, along with marketing. Though it may be taught as a separate course, marketing is part of an integrated company effort to satisfy consumers. At this point, no changes to the marketing strategy have been discussed.

The fourth step is to develop and analyse changes to the previously identified marketing strategy. The key to good decision making is to generate as many alternative courses of action as possible. Some will not be workable, match the environmental constraints, or fully use the capabilities of the company, and can be quickly eliminated. However, three to five major strategic directions should begin to emerge. Evaluating these alternatives to see which best solves the problems or meets the objectives, which were first identified, is critical. One method of analysis is to list the strengths and weaknesses of each option. Consideration should be given to the effect of each alternative on revenue and profitability, meeting consumer needs, and on the other products manufactured by the company.

It is here that learning occurs. Some students may feel that they don't know enough about the industry or a problem to make good decisions. While this may be true, it is important for such students to bear in mind that marketing courses are designed to help them acquire this skill. In analysing an alternative, make certain that the company on which you are focusing is operating from a position of strength. Use the company's strengths to: block threatening actions in the environment; improve weak areas of the company's business; or take advantage of opportunities in the environment. Another common problem in analysis arises when the problem solver tries to do too many things at once. A fire department in a small town cannot hope to extinguish ten fires simultaneously. It chooses the most important building to save. If the fire in that building is extinguished quickly, the department moves to the second most important building. By the next day, seven buildings might be destroyed, but three remain standing. Some of the greatest business blunders have occurred when companies try to take advantage of too many opportunities when their resources are already stretched too thin.

The last step is to make a recommendation and lay out an implementation plan. Based on your analysis, try to recommend one or two alternatives. Sometimes students try to find a way to delay or defer a decision. One delaying tactic is to suggest a search for more information. Such tactics rarely work and can be fatal to an organization. You will need to sort through the alternatives presented and logically support a decision either by showing why your decision is better than the other alternatives or showing why the other alternatives don't solve the situation at hand.

The recommendation is not a conclusion. One does not guarantee the success of a strategy merely by recommending it. It has been said that a brilliant strategy poorly implemented becomes a bad strategy, while a poor strategy brilliantly implemented becomes a great strategy. For the alternative(s) recommended, you should develop a brief tactical implementation plan. Tactical plans list, in order of priority, the specific actions that are required to make the strategy happen. If the recommended strategy is to develop a new product, some of the specific tactics will be to develop a working prototype with the help of research and development, to develop packaging alternatives, to work with an advertising agency to create a series of radio advertisements and to alert retailers to the need for additional shelf space. Timing should also be specifically discussed. It is not good enough to distinguish between short- and long-term tactics; use specific time periods such as a week, a month, or a year. Tactical plans also include control points. In journeying from Toronto to Charlottetown, a traveller not only has a planned route but also intends to be in Quebec City by the end of the first day. If the traveller has not reached the control point of Quebec City after one day's travelling, the plans for the remainder of the trip may need to be altered.

## ◊◊ Some Additional Tips

In a case study, you will be confronted with useful information and misleading information, and some information will be missing. You will need to make some assumptions. While it is not possible to make precise assumptions given your lack of marketing experience, try to state clearly and support the assumptions you are able to make.

Many may think they are unable to make assumptions. How is one to know how much money it takes to introduce a new brand of beer into the market? Is it more than a dollar? Is it less than one billion dollars? If you think both are true, we have now established a lower and upper bound. Granted these limits cross ten orders of magnitude, but they are limits nonetheless.

Maybe we can refine them. Would the cost be more than ten dollars? More than one hundred dollars? More than one thousand dollars? More than ten thousand dollars? More than one hundred thousand dollars? More than one million dollars? Move in the other direction. Less than one hundred million dollars? Less than ten million dollars? From ten orders of magnitude, we have reduced the problem down to one. A likely estimate on the cost of introducing a new brand of beer is between one million and ten million dollars. With a little work, that estimate could be refined further.

Another problem is the excessive repeating and summarizing of the case. It is true that the current marketing strategy, the strengths and weakness of the company, and the opportunities and threats within the environment must be summarized from the case. But a rehash does not demonstrate analytical thinking. Your instructor will not require a summary of the case. Instead, extract only those facts which help lay the foundation for your analysis.

Try to avoid stating a conclusion without providing any reasons for it. Though your argument may seem clear to you, your instructor may see your conclusion as a snap judgement without any connection to earlier evidence. A student could claim, for example, that one weakness of the company is that it is undergoing a profit crisis and lacks funds for investment purposes. The student could then recommend that the company open two additional plants. However, the second idea does not follow from the first. Similarly, some arguments are not brought to their proper conclusion. In such cases, the reader is left in the dark, baffled by dangling statements.

## ◊◊ Written and Oral Communication

When submitting a written analysis, plan on writing two versions of the report. The first is a rough copy. Proofread (don't just use a computer spell-check program) and edit this one. Attempt to do a perfect job of preparing the second copy. While instructors are looking primarily for high-quality solutions, reports submitted with coffee cup rings, small corrections in pen, and no page numbers will not be impressive. It should go without saying that the final version of your report must be double-spaced; have one-inch margins at top, bottom, and sides; and include a title page. Imagine that the report you are submitting is being given to a superior in a company. Would that person be impressed with the look of the report? If it was intended for you, would you be happy with it?

After completing the first version of your report, have other people read it and suggest changes. Have them pay special attention to spelling, grammar, sentence structure, and the flow of logic. Even if they have never taken a marketing course, they should be able to understand the material being presented. You know the report is well-written when these readers ask questions and offer advice on the content being presented.

Avoid generalizations. Some people believe that a *specific* statement is either right or wrong while a generalization is always partly right. This is not so: generalizations weaken your arguments.

Avoid using special marketing terms, especially if you are trying to impress an instructor. You might be tempted to sprinkle your case solution with terms such as *demographics, psychographics,* or *family life-cycle stages*. However, only use these special terms if you know what they mean. If you misuse and misspell them, you will damage your argument.

There is no single correct solution to any case situation. Still, there are many wrong solutions. In class, an instructor will press for a solution that will best enable the decision maker to deal with the problems presented in the case. If you are asked to submit a written report, you will not have the benefit of classroom experience. You can, however, simulate that experience by joining an informal working group. By meeting before class, you can share your ideas and be exposed to other points of view. The result of a group discussion should offer you the chance to explore issues and provide a better understanding of the case situation.

Part of the instructor's task is to facilitate discussion by means of intensive questioning. Through the dynamics of the classroom, ideas can be developed and fully explored by drawing on the different points of view held by different students. Along with helping to manage the class process and to ensure that the class achieves an understanding of the case situation, the instructor helps students develop oral-communication skills and thinking capabilities.

Most courses are designed to build week by week on material already covered and discussed. Class attendance is necessary if learning is to occur under the case method. But while listening to the discussion might give you some insight, it is critical that you become a participant. It is natural to fear talking in front of the class but it is a fear you must overcome. After you graduate, you will find that the people who can present their ideas and convince colleagues and superiors of their merit will advance more easily and quickly within a company.

## ◊◊ Some Final Thoughts

You are encouraged to deal with each case study as it is presented. You should put yourself in the role of the decision maker profiled and look at the situation through her or his eyes. As many of the cases happened in the past, it is tempting to use today's hindsight to make yesterday's decision. Don't do that. Analyse each case using the material available at that time; this way you will confront the dilemma on the same terms as the initial decision maker did. Using hindsight is not an effective way to develop your skills at marketing strategy.

Some students may try to make contact with the company or decision maker. This practice should be discouraged. Companies have kindly donated their time to the creation of a case with no expectation of further commitment. Additional information about a case could also thwart teaching objectives. The facts of the case and its mode of presentation have been chosen to facilitate learning. While it may be necessary to present general information about the period during which the case occurred (for instance, you may be told that there was a recession in 2008-09, or that the Trans-Pacific Partnership was signed in 2015), additional information will not necessarily improve your solution.

Though it may seem daunting, the case method is an enjoyable and stimulating learning tool. Now that you have been briefed, try the first case study assigned. Good luck!!

# MARKETING RESEARCH

You are sitting at home watching television when a commercial break begins. In one commercial, a consumer is confronted with two soft drinks and asked to taste Brand A and then Brand B. When asked which soft drink he or she prefers, the consumer chooses Brand A. Then the narrator adds the tag line: "Research has shown that half the consumers who participated in a similar taste test also chose Brand A."

We have all seen commercials, for a host of products, that have a similar plot line. The interesting question is: What message should someone watching television receive from that commercial? An erroneous conclusion is that Brand A is preferred to Brand B. Given two choices, one would expect people to randomly choose Brand A half the time. The faulty conclusion lies not in the research, but in the inferences drawn from that research.

If the marketing concept involves finding out what people want and giving it to them, then marketing research is of primary importance to that concept. Marketing research *is* the art of finding out what people want. It is an ongoing process of information gathering and analysis. Today that process is known as a marketing information system. But rather than examine an ongoing process, we should try to understand the process of conducting a single research project.

## ◊◊ The Research Process

As is so often the case, the need for a specific research project is triggered by the identification of a problem or market opportunity. (See fig. 1) Care must be taken at this stage to define the research project; it has often been said that a well-defined problem is half-solved. Research should address not the symptoms of the problem but the problem itself. Suppose the triggering event for the research project is a decline in sales. This is just a symptom. The problem could be caused by the actions of a competitor, shifting consumer tastes, currency fluctuations that make cross-border shopping more attractive, changes in tariff rates, and so on. Before a company begins to design a questionnaire or choose a research method, the problem needs to be fully explored and understood.

Part of that understanding leads to the development of some very specific hypotheses, which can be tested through research. Men consume more of the product than women. Older people use the product more often than younger people. People with university educations purchase more of the product than people with high-school educations. These specific hypotheses are critical when it comes to designing an experiment or questionnaire and choosing a research method.

Another critical decision is the type of data that will be gathered to test the hypotheses. Secondary data has been gathered and published by someone else. It could come from internal company records, financial reports, or salesperson meetings, or it could come from such external sources as Statistics Canada, A.C. Nielsen, Environics, Gallup, university research, or government departments. While secondary data can be found in the public domain, a company will generally have to pay a fee for its use. However, that fee is often less than what it would cost the company to gather that information itself. Time can also be saved using secondary data. A research project could take a minimum of ten days to complete, while secondary data could be available in a few hours.

*Figure 1.* THE MARKETING RESEARCH PROCESS

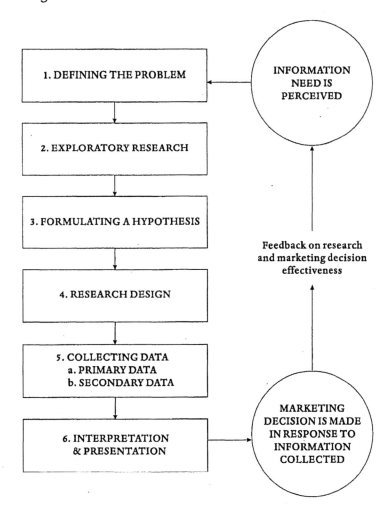

SOURCE: Dale Beckman, David Kurtz, and Louis Boone, *Foundations of Marketing*, 5th ed. (Toronto: Dryden, Holt, Rinehart and Winston of Canada, 1992), 107.

A company may be forced to gather primary data if secondary data is outdated. For example, secondary data on the ownership of smartphones from 2006 is not much use in 2017. Sometimes the method by which secondary data is classified makes it unusable. A company might want to know about the brands of soft drink purchased by those people sixteen to twenty-four. Many studies divide consumers into age groups with ten-year spans – say from ten to nineteen and twenty to twenty-nine. Such categories would not be of much help. The most common reason for undertaking primary-data collection is that secondary data is simply not available. Reaction to a politician's speech on Monday or a change in a pricing policy on Thursday is not available from previously published sources.

## ◊◊ Methods of Data Collection

An experiment is an attempt to establish a cause-and-effect link between variables. Here is a simple experiment: Hold a pen at arm's length and let it go. I guarantee that it will fall to the ground. If we were to take some measurements under controlled conditions, we would determine that the pen was exposed to a constant force that caused the speed with which it fell to increase by 9.86 metres/sec$^2$. We know that force to be gravity. While science deals primarily with inanimate objects like atoms or planets or compounds, marketers deal with people. As people do not always react the same way to the same stimuli, marketing research experiments are much more difficult to undertake. This explains why experimentation is one of the least used methods of gathering marketing information.

Experiments can be conducted in a laboratory setting or in the "real" world. In the laboratory, the experimenter can control all extraneous variables to make sure that variations in the dependent variables are being caused by manipulations of some independent variable or variables. Unfortunately, most consumers do not make decisions or buy products in a laboratory. In the "field," an experimenter has less control over the experiment, but consumers are able to respond in a realistic setting. One of the most often used field experiments is test marketing. In this procedure, a new product or new promotional campaign is introduced to a medium-sized city such as Lethbridge or Guelph. The proposed marketing plan is reproduced on a small scale in the city – from the number of coupons delivered to the shelf facings in the store to the amount and type of advertising used. The marketing plan becomes the independent variable, and the experimenter measures the sales and consumer reaction to the product as the dependent variables. If a product does not generate the desired sales level, or is not well-received by consumers, it is withdrawn from the market. Though a test market could cost several hundred thousand dollars, it is an excellent insurance policy against the millions that national introduction of a product to the market would cost.

A second technique is to view the actions of respondents. Observation studies could include counting the traffic which passes a potential site for a new mall, watching shoppers as they make their way through a grocery store, and monitoring the lengths of the lines at checkout counters. If we observe the actions of consumers, we are not obliged to rely on their capacity to remember or their desire to answer in a socially acceptable manner. Watching the patrons of a bar to see if any drink to excess and then attempt to drive home gives us a much more accurate picture than we would get if we questioned the same people about their behaviour. Observing people may also be the easiest, or the only way to gather information. How could you really determine computer users' responses to a new keyboard design if you did not watch them trying to use it?

A major drawback to observation is the difficulty of interpreting what you have seen. A child picks up a toy, plays with it for thirty seconds, cries out, and then runs to mother. What does that mean? Is there something wrong with the toy? The child? The room? Also, we all bring our biases to what we are seeing. A man in his mid-twenties arrives at a store dressed in an old T-shirt with the sleeves torn off, a pair of jeans with holes in the knees, and a baseball cap worn backwards on his head. Some people would see a man with a low income. Some would see a man who had been working around the house, had used the last bit of some product, and was there to buy more. Others would see a trendily dressed, average student.

The final technique, and the most commonly used, involves the structured answering of questions and recording of answers. Survey research can easily provide the answers to questions such as: Who? What? When? Where? How? As you have gathered, Why? is always the most difficult question for the researcher to answer. In survey research, people can be asked why they do something and they will give a response. But the answer is often meaningless because consumers really don't know why they do something.

There are three primary methods of conducting survey research. The first is to distribute a self-administered questionnaire to respondents to complete at their own pace. The questionnaire may be delivered by hand, mailed, or sent by electronic mail. A major drawback of this method is the response rate. Between twenty-five and fifty percent of self-administered questionnaires are completed and returned. Incentives such as a lottery ticket or a two-dollar coin are sometimes used to improve response rates. A researcher is always concerned that there may be some form of systematic bias that prevents some people from responding. For instance, self-administered surveys require fairly good language skills. People who, for any one of a number of reasons, have problems reading and/or writing in the language of the questionnaire are less likely to participate. Even if people will respond, the questions put to them must be carefully worded. Ask someone where he or she was born, and you might get answers as diverse as "Medicine Hat," "Mexico," "Europe" or "in a hospital."

A second method of conducting survey research is by telephone. When calls are to be made locally, phone surveys can be cheap and quick. If long-distance calls are required, WATS lines (whose numbers typically begin with 1-800) may be used to reduce costs. On the phone, it is difficult to ask long questions – reading someone a list of twenty brands of peanut butter so that she or he can identify the one they purchase most frequently is very tedious. Some people can't easily be reached on the phone – those with unlisted numbers, those who have moved since the phone book was published, and those who do not have a telephone. In many cities, one-quarter of the population is not listed in the telephone directory. Gathering personal information over the phone can also pose difficulties. If a male researcher asks a female respondent about her marital status, her annual income, where she lives, and her hobbies, it is not unusual for her to hang up. People administering a survey over the phone are complete strangers whose motives are always suspect.

The final method of conducting survey research requires an administrator to ask questions and record answers in a face-to-face situation. Personal interviews are a means of obtaining more detail by probing and clarifying a respondent's answers. Because of this detail, personal interviewing is a very slow method of collecting data. Also, the increased human involvement adds substantially to costs. Personal interviews easily accommodate the skipping of questions as well. For instance, if someone has purchased shoe polish in the last month, the researcher might want to ask a question concerning brands or colours. If shoe polish wasn't purchased, the follow-up question might involve a completely different product category.

While most studies attempt to generate quantitative answers that can be analysed by a computer, personal-interview studies are especially useful when qualitative market research is the goal. One qualitative technique is the focused group interview. Here eight to twelve people are assembled in a special room equipped so that the discussion can be either audiotaped or videotaped. Such rooms often have one-way mirrors, which allow an observer to view the session without intruding. With the help of a moderator, the group discusses a topic, in a very non-structured manner, for one to two hours. In the end, the researcher will have developed a better "feel" for the subject being studied.

## ◊◊ Closing the Loop

When a design and method are chosen, the researcher then needs to talk to consumers. In some cases, the number of consumers is so small that each can be contacted. That process is called a **census**. Generally, it is only possible to conduct a census when dealing with industrial customers. For convenience and economic reasons, a subset of the consumer population must be chosen. This sample can be generated in one of two ways. If every member of a population has an equal chance to be chosen, the technique used is called probability sampling. For instance, if the names of all students enrolled in the third-year business program are placed into a drum and one hundred are randomly drawn, every student will have an equal chance of being selected. For a probability sample, the presence of a master list is critical. When such a list is not available, some members of the group in question will, for whatever reason, have no chance of being selected. Consider the task of talking to families who have at least one child under the age of five. No master list of these families exists. To study them, one might visit a day-care centre or a pre-kindergarten and attempt to talk to parents as they drop off their children. However, some families do not or cannot use these facilities and they will have no chance to participate in the study.

Once the questions have been asked and the data gathered, it must be analysed. You probably have some familiarity with statistical computer-software packages and the difficult task of data analysis. Results must be assembled into a written and/or oral presentation. The presentation of the results of statistical analysis is an art that lies beyond the scope of this book.

Creating graphs, charts, tables, and text that both convey information and keep a report interesting is not easy to do. You will no doubt have a chance to practise this art during your course of study.

## ◊◊ The Ethical Dilemma

Though ethics is an important consideration in all aspects of business, it is especially important in research. One of the common complaints about marketing research is that it invades consumer privacy. Most people would not mind being asked how many cups of coffee they had consumed in the last week. Similarly, most would not mind being asked if they had any problems registering for academic classes. But imagine being asked how many bank accounts you have or if you have ever undergone lipo- suction. You would likely say that these questions are an invasion of privacy. Why? The answers to both sets of questions help a company satisfy consumer needs. The questions in the second set, though, touch on subjects that most of society considers too personal. Ethics are, after all, influenced by societal norms of behaviour.

If you intend to offer money to consumers to induce them to complete a questionnaire, should you give it to them before or after they complete the questionnaire? If identifying your client could bias the results, are you under any obligation to tell respondents who that client is? While one requires permission to survey consumers on private property (say in a mall), is it ethically correct to survey people on the public sidewalk just beyond that property? If a questionnaire will take twenty minutes to complete, is there anything wrong with telling a respondent that it will only take five minutes? You may explore the answers to these questions with your class- mates and instructor.

You will encounter three different levels of ethics in your life. These are: individual ethics, institutional ethics, and professional ethics. Individual, or personal ethics constitute the moral code by which we live our lives. This code would cover issues such as premarital sex, abortion, our views on killing, lying, and so on. When you join a company, you will encounter a set of institutional ethics. While this rarely deals with the issues just presented, there will be policies on the rights of co-workers, harassment, proper business practices, environmental pollution, and workplace safety. As well, some career paths might bring you into contact with professional ethics.

Doctors, lawyers, and priests all have codes of ethics that govern their professional behaviour. So, too, do marketing researchers, sales promotion company employees, and those who work for advertising agencies.

Where there is no overlap, there is no conflict. But what happens when you personally abhor environmental pollution and are working for a company that has no corresponding institutional ethic? What happens when your employer asks you to conduct some research in a way that is inconsistent with your professional ethic? You now have a problem. Do you blow the whistle? Do you change your personal code of ethics to keep the job? Ethical dilemmas are never easy to answer and can arise several times a day.

# CONSUMER BEHAVIOUR

When Bill Clinton was elected president of the United States, saxophone sales and lessons skyrocketed. Why? When Roberta Bondar became Canada's first woman in space, why did enrollment by women in university science and engineering programs increase? When the television series "The Big Bang Theory" became a hit, stars Johnny Galecki and Jim Parsons set a fashion trend with their superhero T-shirts. Why?

The answer lies in psychology and sociology. If companies are waging a battle for consumers, the battleground is the mind. Many marketing concepts have been borrowed from these disciplines. Quantitative marketing research can easily characterize consumers; it can determine when and where they shop, what they purchase, and how often they shop. A good working knowledge of psychology and sociology can help us understand why consumers buy.

In answering the questions posed earlier, we may find that the psychological concept of self is helpful. There are three aspects of self. The first is our real self: the objective view of oneself as one really is. The second is the looking-glass self, which is one's interpretation of how one is seen by others. Finally, there is the ideal self, which represents the way one would like to be. Parts of all three selves combine to form self-image – the way one sees oneself. Why is this important to marketers? Because we know that people buy products that move them closer to their ideal selves.

Consumer behaviour is the series of actions in which individuals engage to obtain and use goods and services. Decision processes are influenced by two major factors: personal influences, which are self- generated; and interpersonal influences, which come from family and friends.

## ◊◊ Personal Influences on Consumer Behaviour

Our purchasing actions can be traced to four theoretical concepts which have been inferred from psychological studies (see fig. 1).

Purchases begin with a need, a lack of something useful. That lack must be so powerful that it impels you to satisfy it. For instance, you lack a Barbie doll. That lack does not, however, have the power to make you do anything about it. You don't **need** a Barbie doll. Motives are intimately linked to needs. They are internal mental states that direct us toward satisfying a felt need. Suppose you have a need for food (a very powerful lack). Motivators might include a growling stomach, a feeling of weakness, a headache, or a sudden craving for a chocolate-chip cookie.

Abraham Maslow proposed a hierarchy of needs based on two premises: our needs are based on what we already possess; and once one need has been almost satisfied, another will emerge (see fig. 2). The most basic needs are physiological: food, shelter and clothing. In some developing countries, people spend most of their lives trying to satisfy these needs. Once satisfied, a series of safety needs emerge: protection from physical harm and shelter from the unexpected. In places where guerilla warfare rages, people struggle daily for security and safety. At the third level, we face social needs; we must belong to, and be accepted by, groups and family. In Maslow's original writings, he claimed that most North Americans were able to reach this level of the hierarchy, but few could progress beyond it.

*Figure 1.*   PERSONAL INFLUENCES ON CONSUMER BEHAVIOUR

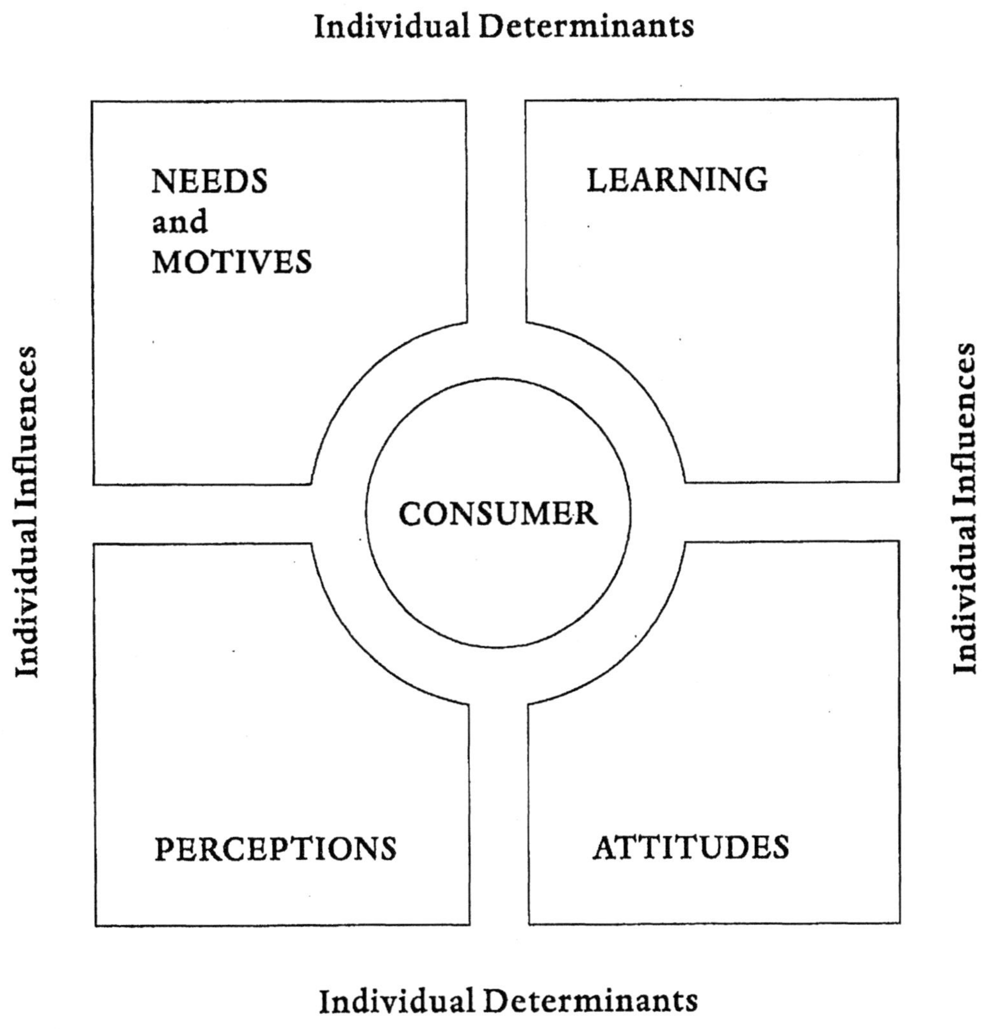

**Individual Determinants**

**Individual Influences**

**NEEDS and MOTIVES**

**LEARNING**

**CONSUMER**

**PERCEPTIONS**

**ATTITUDES**

**Individual Influences**

**Individual Determinants**

SOURCE: Adapted from C. Glenn Walters and Gordon W. Paul, *Consumer Behaviour: An Integrated Framework* (Homewood, IL: Irwin, 1970), 14.

Near the top of the hierarchy are esteem needs: status, praise, recognition by one's peers. Though linked to social needs, esteem needs arise from the desire for respect from groups or family members. Maslow suggested that only a few people reach the top of the hierarchy, where self-actualization needs are situated. Self-actualization refers to a feeling of using one's talents and capabilities totally, of realizing one's full potential. Perhaps people such as the Pope, Mother Teresa, Martin Luther King, Albert Schweitzer, and Bishop Desmond Tutu have been able to reach a state of self-actualization.

A couple of problems arise in applying Maslow's hierarchy. First, there is a North American/Western European cultural bias. In Southern and South-east Asia, social needs may come before both safety and physiological needs. Second, Maslow's hierarchy applies to groups of people and not individuals. I doubt that consumers in a supermarket will tell you that they purchased a jar of Kraft peanut butter because it fulfilled their social needs.

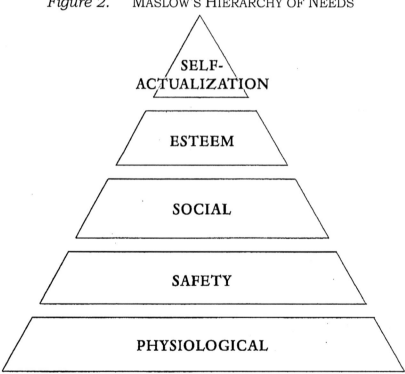

*Figure 2.*   MASLOW'S HIERARCHY OF NEEDS

SOURCE: Adapted from A.H. Maslow, "A Theory of Human Motivation," in *Motivation and Personality*, 2d ed. (New York: Harper & Row, 1970).

A second set of influences is perceptions – the meaning we attribute to stimuli received through the five senses (touch, taste, sight, smell, and sound). From psychological studies, we know that two people confronted with the same sets of stimuli will interpret them differently. Just ask police interviewing eye witnesses to a crime. Descriptions of the perpetrator vary widely, even when it comes to aspects such as gender and dress. Hence, as marketers, we need to know that a consumer's perception becomes his or her reality.

As you read this book, you are being bombarded by dozens of stimuli – the sound of air moving, the drip of a faucet, the rustle of your clothes, an aftertaste from breakfast, music off in the distance. If your brain allowed you to be conscious of every stimulus received, you would soon go crazy. As a defense mechanism, you develop filters, or screens, to block out background noise. Only unusual stimuli will break through the screen. This concept can be applied in advertising. Marketers know that people watch nearly an hour of commercials during an evening of television, yet the next day they will consciously remember only a handful of ads. These are the ones which were most unusual, made them laugh or in some way touched a nerve.

## ◊◊ Attitudes and Learning

A third set of influences is attitudes – the favourable or unfavourable feelings, actions, and knowledge – we have toward an object or idea. We cannot measure attitudes directly. The question "Do you have an attitude about ice cream?" does not elicit usable answers on a survey form. Instead, we measure three components of an attitude. Think of them as the ABC's of attitudes: affections, behaviours, and cognitions (see fig. 3).

Affections are your subjective emotions or feelings about an object or idea. Consider the snake. Most people *feel* that snakes are dangerous and slimy. They don't like them. People don't think of snakes as friendly or playful. Affections toward them are generally negative. Cognitions are the objective facts and knowledge you have about an object or idea. All but one variety of Canadian snakes is non-poisonous. They eat insects, frogs, grasshoppers, and rodents. Their skin is scaly and dry. They grow to be about half a metre to a metre in length, and are usually a dark brown or green colour. Objectively, cognitions are generally positive for snakes.

Behaviours represent our tendencies to act in a certain way when confronted with an object or idea. If I threw you a snake, you would probably shout or scream and then run away. This is not a positive behaviour. Though it is an over-simplification to simply add the three components to

*Figure 3.* THREE COMPONENTS OF ATTITUDES

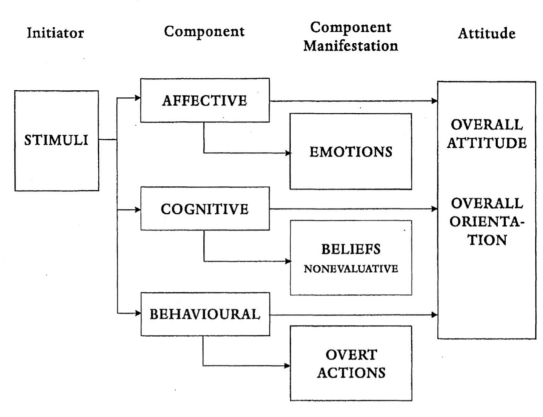

SOURCE: Del I. Hawkins, Kenneth A. Coney, and Roger J. Best, *Consumer Behaviour: Implications for Marketing Strategy* (Dallas, TX: Business Publications, 1980), 334. The figure is adapted from M.J. Rosenberg and C.I. Hovland, *Attitude Organization and Change* (New Haven, CT: Yale University Press, 1960), 3.

determine an overall attitude, from this analysis it is easy to see why society in general has a negative attitude toward snakes.

A marketer's worst fear is that consumers will develop a negative attitude toward a product. When that happens, it is always easier to change the product to match prevailing attitudes than it is to change the prevailing attitudes. In my example, snakes cannot easily be changed so you would have to try to change attitudes. This is a difficult task. There are only three courses of action to try, and these mirror the three components of attitudes.

For instance, if cognitions are negative, new, positive cognitions could be introduced to educate consumers. If affections are negative, the product could be linked to something emotionally positive. Hence, the reason for so many products being linked to physical attraction: sex sells.  Finally, if behaviours are negative, consumers could be asked to engage in attitude-discrepant behaviour.  By distributing free samples or coupons, or by doing blind taste tests, consumers would be encouraged to try the product, even those who say they don't like it.

A final personal influence is learning, which is defined as changes in behaviour as a result of experience.  The learning process begins with a motivator or drive which impels some action.  Cues are objects in the environment which help determine a reaction to the drive.  These cues might be advertisements, shelf positions in a store, billboards, or free samples.  The drive and cues create a response that, for a marketer, is generally the purchase of a product.  This successful combination then needs to be reinforced positively so that it occurs again when the same set of drives and cues are present.  Of course, a response could be a refusal to purchase the product.  In that case, reinforcement is designed to show this as an incorrect behaviour and to stimulate a different response when the same drives and cues occur.  As the process of drive, cue, response and reinforcement is repeated again and again, the response becomes automatic.  This model is really better suited to the training of animals.  It neglects the effect of human free will in the decision process.  Nonetheless, it gives insight into the learning process.

## ◊◊ Interpersonal Influences on Consumer Behaviour

In much the same way as the four P's of marketing operate within a business environment, personal influences on consumer behaviour operate within a sociocultural environment that can also shape behaviour.  Three major factors come into play here (see fig. 4).

The group with the strongest influence on behaviour is the family.  It shapes our behaviour when we are young in ways that do not manifest themselves until adulthood.  As children, our influence on purchase decisions is limited, yet when our parents become elderly, we may be asked to take responsibility for the decisions that affect them.  Families have changed much in the last seventy years.  In the 1940's, it was common to have three generations living under one roof.  Today, both the old and the young seek independent living alternatives.  The divorce rate of over fifty percent has triggered an increase in single-parent households.  Men are being awarded child-custody rights, which were once given almost exclusively to women. More women are working, yet the average work week has shrunk to less than forty hours. Men are more likely to be involved in shopping, though they still do less of it than women.

*Figure 4.* INTERPERSONAL INFLUENCES ON CONSUMER BEHAVIOUR

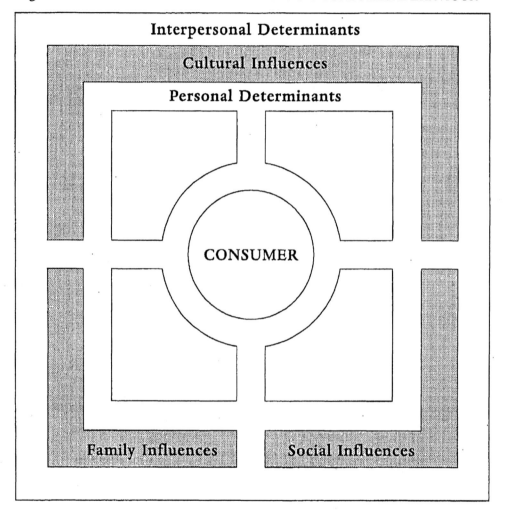

SOURCE: Adapted from C. Glenn Walters and Gordon W. Paul, *Consumer Behaviour: An Integrated Framework* (Homewood, IL: Irwin, 1970), 16.

Married heterosexual couples can make decisions about which products to purchase in four different ways. Decisions can be male dominant (for example, the decision to purchase a car or insurance policy) or female dominant (the purchase of food, kitchenware, or children's clothing). They can also be made jointly (the purchase of a home, vacation, or furniture).

Finally, decisions can be made equally by either the husband or wife (the purchase of milk or non-prescription drugs whenever they are required).

A second important influence is social groups. From these groups, we acquire both status (a position in relation to others) and a role (expectations of others about our actions). In class, your instructor has a specific status and you expect her or him to act in a certain way. As students, you are subject to behavioural expectations as well. Groups that can influence behaviour are called reference groups. You do not need to formally join such a group for it to have an influence on you. You might choose to dress like a racing cyclist in lycra shorts and aerodynamic shirt even if you don't own a bicycle or belong to a cycling club.

Reference groups can influence you to purchase a product or a specific brand. For instance, a particular reference group might influence you to buy a telescope or a motorcycle. It might influence your choice of magazines or music. We also know that this influence is quite strong. People like to conform to the norms that are established by group leaders. Within a group of cyclists, there is probably one who is the first to test each technological advance and then either encourage or dissuade others in the group from making a purchase. It is important for marketers to identify these opinion leaders and to gauge quickly their reactions to new products.

A final interpersonal influence is culture, which includes the values, beliefs, attitudes, and institutions created by a group of people. These values and institutions shape human behaviour not only today but also in generations to come. In Canada, which has a rich cultural mosaic, there are many subcultures that represent opportunities for the marketer. The largest subculture is that of French Canadians. We know that this group has a sweeter tooth, has a greater joy of food and drink, and is much more family-centred. Subcultural groups can be defined by ethnicity (Italian, Pakistani, Ukrainian), religion (Catholic, Jewish, Islamic), and social group (skinhead, preppy, biker). A uniquely Canadian government policy encourages sub-cultures to retain their values and institutions rather than be assimilated into the general Canadian culture.

In an internationally context, cultural differences are very significant. Chevrolet's plans to introduce the Nova to Latin America came to an abrupt halt when someone finally realized that in Spanish "no va" means "won't go." In an Arabic country, an appliance manufacturer displayed a billboard showing dirty clothes, a washing machine, and clean clothes – a clear implication that the washing machine worked well. However, Arabs read from right to left. The implication of the ad in the Arab world was that the washing machine was good at making clean clothes dirty. In parts of Europe, the main gift-giving occasion is not Christmas Day (December 25); gifts are presented at varying times throughout January. Clearly, a marketing program that works well in one country or with one culture will not necessarily do the same in another setting.

## ◊◊ The Consumer Decision Process

As demonstrated by figure 5, the consumer decision is the series of steps that a consumer takes in determining a need for a product and choosing the brand to purchase. Both personal and interpersonal influences affect the consumer at every step of the process.

The process begins with the recognition of a need. It may be that a person has consumed her or his supply of a product or is bored with the brands on hand. It may be that a consumer has had a bad experience with a brand or simply wants a greater variety of options. A consumer might have come into some money, become unemployed, or recently married, and is thus looking for different types of products. With a recognized need, consumers begin to search for background information to help them make a decision. Searches can be internal (recalling what he or she already knows) or external (visiting libraries, talking to others, obtaining brochures). A search can be fast (the micro-second firing of a synapse) or slow (many years) but when it is completed, the consumer has determined a set of products or brands from which to make a final choice.

This set of brands is then logically evaluated. This evaluation may also vary in duration. The criteria used may be objective (for example, the price, colour, or size of the object), or subjective (the reputation, flawlessness, or feel of the product). While it may appear that dozens of criteria are required to purchase a complex product such as a house or car, studies show that most consumers use at most seven criteria to guide their decision making. With multiple criteria, the results on one dimension may be used to compensate for deficits on another dimension. For instance, a consumer choosing between two cars may prefer the black one but discovers that the yellow car is twenty-five percent cheaper. In other cases, no compensation is allowed –

the consumer only wants a black car, regardless of its rating on other criteria. Consumers make decisions using their own models and judgement, but at the end of this stage a definite purchase decision has been made.

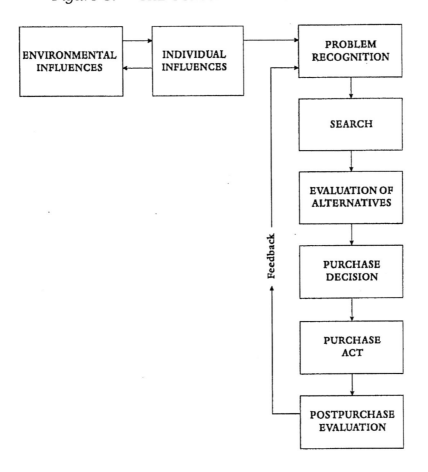

*Figure 5.* THE CONSUMER DECISION PROCESS

SOURCE: Adapted from C. Glenn Walters and Gordon W. Paul, *Consumer Behaviour: An Integrated Framework* (Homewood, IL: Irwin, 1970), 18; and John Dewey, *How We Think* (Boston: C.C. Heath, 1910), 101–05. Similar steps are also discussed in Del I. Hawkins, Roger J. Best, and Kenneth A. Coney, *Consumer Behaviour: Implications for Marketing Strategy*, rev. ed. (Plano, TX: Business Publications, 1983), 447–606.

The next stage involves the purchase act. The product purchased may or may not be identical to that decided upon earlier. For example, you may have decided to purchase a bag of Oreo cookies, but when you get to the store the shelf is bare. Though you decided you wanted Oreos, you may leave the store with Fudgee-O's.

Having completed the purchase, you may feel some anxiety. Did I buy the right product? Did I pay too much? This psychologically unpleasant state is called cognitive dissonance. It is most likely to occur when the decision is major (choosing a university to attend), represents the expenditure of many dollars (buying a house), or when the rejected alternatives are not clearly inferior to the item purchased. The marketer from whom you purchased the product will want to help you reduce the dissonance by providing personal reassurance and by sharing supportive information. Of course, a marketer representing the competition will attempt to increase dissonance so that you regret your decision and will consider making a different decision, should a similar need arise. The latter was the premise behind the immensely successful "Wow! I could have had a V-8" advertising campaign.

# INTERNATIONAL MARKETING

The North American Free Trade Agreement. The European Union. The Trans-Pacific Partnership. The General Agreement on Tariffs and Trade. The World Bank. The International Monetary Fund. The G-8 countries. Unless you have been hibernating for the last decade, these names should be instantly recognizable. While few of us know, in detail, what they mean, we realize that trade and marketing are becoming an increasingly prominent global concern.

It has not always been thus. Seventy years ago, few countries were able to produce enough goods to satisfy internal demand. Transportation options were few and fairly unreliable. The world standard for currency was gold and that was difficult to carry.  Communications with home offices took days, so it was not possible to check the credit worthiness or legitimacy of a buyer from another country.  In those early days of inter- national trade, crooked dealings and illegal activities were quite common.

Perhaps that, in part, explains the sluggishness with which Canada and the United States have pursued international trade.  As North America was one of the first areas to develop substantial surpluses, it was also one of the first to suffer through the problems and pitfalls of international trade.

But the technological advances of the last few decades have closed the loopholes and made international trade more possible and desirable and less difficult than ever before.  Increasing affluence around the world, the need for technology transfer to developing countries, mass and instantaneous access to communication media, an unprecedented period of world peace, and a rapid realization that Earth is a much smaller planet than first imagined have brought the world's markets closer together.

## ◊◊ Some Terminology

When a country produces goods for sale outside of its domestic market, it is engaging in exporting.  Importing occurs when a country purchases goods from outside its borders for use internally.  For every billion dollars of exports, a country can generate thirty thousand to forty thousand jobs.  Not surprisingly, a dependence on imports can cause an increase in national unemployment.

The Balance of Trade may be thought of as an interesting relationship between exports and imports.  If exports exceed imports, a country has a trade surplus or a favourable balance of trade.  When imports exceed exports, a country has a trade deficit or an unfavourable balance of trade.  For most of the last decade, Canada has had a trade surplus while the United States has had a trade deficit.

Another measure looks at flow of cash between nations.  Starting with the balance of trade, one adds or subtracts spending on tourism, the military, foreign aid, and foreign investment to arrive at the balance of payments.  Primarily because of the interest payments on the national debt that Canada must make to foreign countries, and because Canadians love travelling outside the country, Canada has an unfavourable Balance of Payments.  In essence, this means that cash is leaving the country.

Because business and financial transactions take place around the world, exchange rates are a major consideration. These are the rates at which one nation's currency can be exchanged for other currencies. Francois Mitterand, the late President of France, once performed a startling experiment. He had the equivalent of one hundred dollars in French francs converted to local currency as he toured the twelve founding nations of the European Union. When he returned to France, he had less than half the money with which he started, yet he had made no purchases. Mitterand used the results of this exchange-rate experiment to help convince members of the Union to adopt a Euro dollar as a common currency.

## ◊◊ Trade Restrictions

The simplest trade restriction is called a tariff. It is a tax levied against imported products. Tariffs or duties may be imposed by a country as a source of revenue. Tariffs on tobacco products in Canada are mainly levied to raise money for the federal government. Tariffs may also be imposed to protect domestic products from cheaper imports. As Canada does not produce cotton, textile manufacturers face higher costs and thus produce higher-priced goods. The government has imposed a tariff on imported textiles to protect Canadian manufacturers and improve their chances of economic survival. Tariffs may be levied as a fixed amount (fifty cents per item) or as a percentage of the value of the product.

The General Agreement on Tariffs and Trade was established internationally to encourage freer trade and reduce tariffs. This agreement makes levying either a new revenue or protective tariff very difficult. To be allowed to levy a new revenue tariff, a country has to show economic need. To levy a new protective tariff, a country has to demonstrate that competitive harm is being done to its domestic industries. The most often cited example of harm is dumping, or the selling of products at significantly lower prices in foreign markets than in a country's domestic market. In recent years, Canada has been accused of dumping cedar shakes and shingles, steel, beef and lumber. In turn, Canada has claimed to be the victim of other countries, which have dumped toothpicks, shoes, steel, computer chips, and wine.

If competitive harm is so great that tariffs have no effect, a country may impose an import quota. By setting limits on the volume of a product which may be imported, the country hopes to minimize damage to its domestic industries. In the 1980's, sales of Japanese automobiles grew dramatically in North America. To protect national car manufacturers, steep tariffs were imposed often increasing the price of a Japanese automobile by over one thousand dollars. This had little effect on sales. Both Canada and the United States imposed import quotas on the number of Japanese automobiles brought into North America. Such restrictions are not in place today, as Japanese car makers responded by building plants in North America and assembling the cars here, thus protecting the jobs of Canadian and American auto workers.

The most extreme form of an import quota is an embargo. This is a total ban on importation of a product, and is most likely to result from political or legal, rather than economic, pressure. To force governments to change their policies, in the last fifty years, Canada has imposed embargoes on all products from South Africa, Iraq, Iran, Syria, and Bosnia. If sanctions are to be fully effective, all nations have to participate. The United States imposed an embargo on trade with Cuba in the early 1960's, yet Cuba survived by trading with Communist countries and others, such as Canada, that did not join the United States in its action.

## ◊◊ Levels of Involvement in International Marketing

There are five levels at which a company can be involved in international marketing: casual exporting, active exporting, sales offices or branches, foreign licensing, and foreign production and selling. Casual exporting is the lowest and most passive level. In essence, it is what happens when a company exports goods only when it has a surplus or, in some cases, without even meaning to export product. For instance, a visitor from Italy may come to Canada and develop a fondness for maple syrup manufactured in Quebec. That visitor may then decide to purchase, through a relative, four litres of maple syrup each year and have them shipped by that family member to Italy. The manufacturer will have no knowledge of the export. Clearly, casual exporting requires no sustained effort on the part of a manufacturer.

When a firm is prepared to make an ongoing commitment to selling its merchandise in other countries, it is involved in active exporting. The ongoing commitment can vary from regularly filling a few orders from one country to setting up a structure to process orders from around the world. For most Canadian firms interested in international marketing, this level represents the extent of their commitment.

Once demand for a company's product outside the country reaches a certain level, it may make economic sense for that company to become more committed to international markets by establishing a sales office or sales branch in another country. A sales branch collects orders through salespeople and fills those orders from stock held on site. A sales office does not carry stock; orders are filled from stock held in Canada.

Another level of international involvement is attained when a company shifts some production from its home base to a foreign country. This may be accomplished, for example, through licensing. This involves a domestic firm allowing a foreign company to produce and distribute its products. In exchange for the production technology, the licensee agrees, in a formal contract, to pay a royalty – usually some combination of a lump-sum payment and a percentage of sales revenue. The licensee also contributes its knowledge of local markets, distribution channels, local laws, and advertising media. Nearly half of all licensing agreements end in failure. It is possible for the licensee to learn about the technology, improve upon it in some way, and then cease paying royalties. For that reason, technology that is nearing the end of its useful life span is more likely to be licensed. Licensors may also err by not providing full technical help to overcome production headaches that arise when there are variations between the raw materials used by the licensee and those used by the licensor itself, or when there are problems with environmental conditions.

By combining out-of-country manufacturing with out-of-country selling, a company reaches the highest level of international marketing involvement. A firm can engage in foreign production and selling by building its own facilities in another country, by buying an existing foreign operation and adapting it, or by entering into a joint venture with another company. In a joint venture, a company joins with a foreign counterpart to create a new company jointly financed and managed.

While few companies reach this final level, multinational companies are becoming more important in global markets. These companies view the entire world as their market, and can service this market through their plants located in many different countries. Coca-Cola, General Motors, Nestlé, McDonald's Restaurants and Disney are just a few of the multinationals at work in the world today.

## ◊◊ Multinational Economic Integration

In the last forty years, the world seems to have become a smaller place. With instant communication links, exposure to mass-media coverage, access to travel, and better access to information, the peoples of the world know much more about each other's histories, cultures, and economies. Seventy years ago, the world saw political alliances designed to either increase or balance military power. Today, rapidly forming economic alliances prevail. Clearly, few countries have access to the raw materials and technology required to produce enough goods and services to satisfy all their internal needs. The need to trade is a starting point for integration.

One of the simplest forms of economic integration is free trade, whereby participants agree to remove tariffs on the goods they trade among themselves. A free-trade agreement may not cover all goods produced, nor will it be implemented instantaneously. Normally, there is a phase-in period, which gives firms a chance to adjust to the new trading opportunities and threats. The North American Free Trade Agreement and the Trans-Pacific Partnership are two examples of this form of economic integration. Rather than forging a wide-ranging agreement, some countries establish free-trade zones. In these zones, products may be imported with no duty. Often these products are component parts which are assembled into finished goods destined for other countries. Tariffs are paid only once: when the finished good enters the country that is its final destination. The establishment of free-trade zones can bring jobs and investment, possibilities that, if tariffs had been imposed, might not have occurred.

A stronger form of economic integration is a customs union. This has all the elements of a free-trade agreement but adds a uniform tariff policy for trade with non-member nations. When the European Union was formed, the member nations sought to establish a uniform tariff structure so that free-trade gains would not be undermined. Suppose there was no uniform tariff policy. Countries A and B have a free-trade agreement. Country A imposes a tariff of ten percent while Country B imposes a tariff of twenty percent. Outside manufacturers will ship products to Country A, and then use the free-trade agreement to move the product to Country B, thus avoiding the higher tariff. Unfairly, Country A will reap most of the economic benefits of trade with non-member nations.

# INTERNATIONAL MARKETING

The strongest form of economic integration is an economic union. It has all the features of a customs union, but includes additional agreements that allow free flow of capital, services, and workers. Many people have suggested that the North American Free Trade Agreement should be expanded to an economic union with ancillary agreements concerning pollution standards, workers' rights, rules on sexual and racial discrimination, minimum-wage standards, and so on struck between Canada, Mexico and the United States. In 1992, the European Union was formed between twelve member nations (Belgium, Denmark, France, Germany, Great Britain, Greece, Ireland, Italy, Luxembourg, Portugal, The Netherlands and Spain) - an amazing feat considering that many of these countries were at war only fifty years prior.

## ◊◊ Global versus International Concepts

Recognizing the trend toward the formation of multinational companies and toward closer multinational economic integration, educators are placing a greater emphasis on teaching marketing from an international perspective. But what does this mean?

In teaching the concept of cost-plus pricing, an instructor could internationalize the discussion by citing costs in rupees or yuan, adding an appropriate mark-up, and determining the final selling price for India or China. But is this internationalization? No. The concept of cost-plus pricing is global, and can be universally applied. Changing the currency does not really change the application of the concept.

However, as marketing students, you have probably been exposed to Maslow's hierarchy of needs. Let me recap the first three levels. The most basic level involves physiological needs: food, water, shelter, and clothing. On the second level are safety needs: protection from physical harm, protection from the unexpected. Finally, on the third level are social needs: the need to be accepted by family and friends. Is this a global concept? No. The hierarchy outlined applies best to North America and Western Europe. In much of the Southern and South-East Asia, social needs precede safety needs. Some people have argued that social needs even precede physiological needs there. This is an example of an international concept – one that varies from one culture to another.

It is difficult to estimate how much of marketing is global rather than international. Some people believe the split is eighty percent to twenty percent. Some feel it is just the opposite. As a student, you may, at first, find this confusing. But you should see it as a challenge. You should always be asking yourself and your instructor if the concepts you are being taught have global acceptance or if there are subtle variations of which you should be aware.

# EAST HAMILTON MINIATURE GOLF

In January 2014, Jean-Guy Gauthier and Sonia Hamdani, two high school teachers, had just finished a proposal for a miniature golf course in Hamilton, Ontario. The idea developed after watching a miniature golf tournament on a cable sports network. They were discussing whether they should invest the time and money to make the proposal a reality. Jean-Guy felt the proposal would make money no matter where they located or how they promoted the venture.

"Sonia, there's not much primary competition and there are lots of people who would love to play miniature golf in Hamilton. I think we've got a potential gold mine on our hands. I've calculated that our maximum capacity for the course is 864 rounds per day, based on the assumption that there would be four people per hole and they would take one hour to play one round. Given that there are eighteen holes and the course will be open twelve hours per day, a total of (4 x 1 x 18 x 12) 864 rounds could be played every day."

Sonia was more cautious. "I think there are two important factors. First, if we don't get the Centre on Barton location, I wouldn't be too keen on the idea. Second, if we don't promote miniature golf properly, there's a chance that it won't succeed. I think we should have another look at our analysis and determine if this idea could work."

## The Idea

Jean-Guy Gauthier and Sonia Hamdani often discussed ways of getting into business during their lunch hours at school. The months of July and August were relatively slow times for them at their work and they both liked the idea of earning a second income to supplement their base pay. The two teachers felt they could each invest $10,000 in a business venture if they could come up with a reasonable idea. After seeing the televised miniature golf tournament, they decided to do some research on miniature golf in Hamilton. The research included an analysis of competition, potential locations, consumers' needs, the Hamilton market, and the costs involved.

## The Competition

YellowPages.ca revealed eight miniature golf courses in Burlington and Hamilton. **Wedgewood Golf Centre** was not analyzed as its target market was mostly Burlington. **Rock Chapel Golf Centre** was not analyzed as its target market was mostly Waterdown and Dundas. **Eagle Classic Golf Centre, Satellite Golf Centre,** and **Pros Golf Centre** were, at best, secondary competitors as they served people who lived on Hamilton Mountain or in Ancaster. That left three courses as primary competitors: **Adventure Village** (outdoor course connected to a water park)**, Putting Edge** (indoor course with glow-in-the-dark lighting) and **Glover Golf Driving Range** (outdoor course connected to a golf driving range).

The competitors were evaluated on a number of criteria (Exhibit 1), and the general conclusion reached by the partners was that the courses were either of poor quality or were too far away from the proposed location to pose much of a threat. It was felt that if a miniature golf course was constructed of high-quality materials and offered a fair degree of challenge, it would attract most of the competitions' customers. If they went ahead with the venture, the partners wanted to construct the best possible course in terms of challenge, materials, and craftsmanship.

## Potential Locations

After looking at a number of areas, the partners concluded that any location should be readily accessible to the public. The basic idea was to "bring the game to the people" by having a convenient location in or near a shopping centre that had high traffic flows. The manager of Triovest Realty Advisors Inc., the company that operated the Centre on Barton, was contacted, and the idea of a miniature golf course located at the centre was discussed. The Centre on Barton was considered an ideal site as it was the most central of the "big box" shopping areas in the lower City of Hamilton with over 61 stores and services. Major tenants included Canadian Tire, Metro, Marshall's, The Brick, Wal-Mart, Michael's, Staples and Dollarama. Just 100 metres to the east was located East Hamilton Radio, a popular electronics store which drew customers from Niagara Falls to Toronto.

The number of visits to the Centre on Barton each month was estimated at around 340,000. The primary market of 136,685 patrons for the centre lived within a 15-minute commute. The secondary market of 159,250 patrons for the centre lived within a 30-minute commute. The centre had large areas of ground parking space (parking was available for 4,000 cars).

It was proposed that the miniature golf course be located in the parking lot near one of the anchor stores (Dollarama) for the period from May 1 to September 30. For the rest of the year, the wooden miniature golf course would be placed in storage. The manager, while interested in the proposal, did not commit himself to the venture. He suggested that the two partners return after they had finalized their plans. If they were allowed to locate at the Centre on Barton, the rental fee for the land would be 15% of gross sales.

## Consumer Analysis

The next step in the project was to conduct a consumer analysis. The partners listed a number of consumer needs they felt miniature golf could satisfy and ranked them in terms of probable importance for three different consumer groups. The needs and rankings were:

| Consumer Need | Pre-teen & Teen | Male Adults | Female Adults |
|---|---|---|---|
| Recreation/enjoyment | 1 | 1 | 1 |
| Challenge/competition | 2 | 2 | 5 |
| Socializing with friends | 3 | 8 | 3 |
| Family outing | 4 | 4 | 2 |
| Relaxation | 5 | 5 | 4 |
| Status | 6 | 7 | 7 |
| Convenience | 7 | 3 | 8 |
| Time available to play | 8 | 6 | 6 |

This analysis indicated the primary needs satisfied would be enjoyment, challenge, and socializing with friends or family. Further information was collected through two consumer surveys. The first survey, shown in Exhibit 2, asked 100 adults if they would patronize a miniature golf course at the Centre on Barton. The results indicated that consumers might participate in miniature golf while shopping there.

The second questionnaire was designed and given to students at their high school. The results, shown in Exhibit 3, indicated that most students would play miniature golf at the Centre on Barton if they were already there. Approximately 50% would play miniature golf on a date, and 50% said they would come to the centre on Sunday and play. Approximately 77% said they felt that $8.00 was a reasonable price for mini-golf. Only 17% felt that $8.00 was too high a price.

**The Hamilton Market**

In 2012, Hamilton was Canada's ninth largest Census Metropolitan Area (which included Hamilton, Burlington, and Grimsby) with 761,346 people and 299,428 households. It was traditionally a heavy-industry community with large steel mills (US Steel and ArcelorMittal Dofasco) and manufacturing facilities using steel in their product lines. By 2014, seven of Hamilton's top ten employers were in the greater public sector including Hamilton Health Sciences, St. Joseph's Hospital, McMaster University, Hamilton-Wentworth Board of Education, Hamilton- Wentworth Separate Board of Education, Province of Ontario, and City of Hamilton. Mohawk College and the Government of Canada were also significant employers.

In 2012, the average annual pre-tax household income in the CMA was $90,500 compared to the Canadian average of $85,795. On a per capita basis, average annual pre-tax income in the CMA was $35,593 compared to the Canadian average of $34,352. (See Exhibit 4 for more age and income data).

In terms of weather, the average number of days with and without rain between May and September is presented in Exhibit 4. On average, there were 104 days without rain during the planned five operating months.

**Cost Estimates**

The partners calculated the costs of constructing the miniature golf course out of wood (Exhibit 5). The total estimated capital cost of $34,068.70 included the cost of building the eighteen holes plus a pro shop, fencing, and miscellaneous expenses. No budget was included for labour because the holes could be built by the industrial arts class at the high school where they taught. The only operating expenses would be advertising and hiring someone to operate the course. The cost of hiring student labour was estimated at $14,783, based on paying them $10.30 per hour (this was the new minimum wage for students being introduced on June 1, 2014), twelve hours per day for a season of 104 days. (If it rained, they planned that the students would not be asked to work and they would not be paid.) It also included a 15% premium to cover the required benefits (CPP, EI, vacation pay). The owners had planned to have the course operate from 10:00 a.m. to 10:00 p.m. each day.

While they had collected some data on advertising rates, they had not decided on any advertising campaign. The Hamilton Spectator, the local daily newspaper, had a city-wide circulation of 209,000 households. The cost of advertising for a full page, half-page, quarter-page, and one-eighth page was $23,750, $12,000, $6,000, and $3,000, respectively before any discounts for volume purchases of advertising. Hamilton was served by eight radio stations. CIOI and CFMU were based at Mohawk College and McMaster University respectively and were not available for standard commercial advertising. CHML, CHAM, and CKOC were AM stations while CHKX (New Country 94.7), CJXY (Y108 Classic Rock), CING (95.3 Fresh Radio) and CKLH (K-Lite) were FM stations. Radio advertising costs ranged from $25 for a thirty-second prime time spot on 95.3 Fresh (adult contemporary) to $250 for an equivalent spot on New Country 94.7 or K-Lite (continuous light hits). Discounts for volume purchasing of advertising could reduce these rates by up to 50%.

**Decisions**

The partners faced a number of decisions. They had not decided the final price to charge, either $8.00 or $10.00 per round; what advertising should be done, if any; or what they should do if the manager of the Centre on Barton did not agree to their proposal. They estimated the total capital cost would probably be around $34,069. A colleague in the Business Department recommended that they start the business with cash to cover two months of operating expenses (wages and advertising). They felt this would mean borrowing $15,000 to $25,000 from the bank. Finally, the major decision had to be made. Should they invest in this venture?

**Exhibit 1  Primary Competitor Analysis**

|  | Adventure Village | Putting Edge | Glover Golf Driving Range |
|---|---|---|---|
| Location – Accessibility | Excellent | Excellent | Good |
| – Built-in clientele | Excellent | Good | Fair |
| Cost – Per 18-hole round |  |  |  |
| Adult | $11.00 | $12.00 | $6.00 |
| Student | $11.00 | N/A | $5.00 |
| Children | $9.00 | $9.00 | $4.00 |
| Course – Appearance | Excellent | Excellent | Fair |
| – Challenge Offered | Very Good | Excellent | Fair |
| – Material Quality | Very Good | Very Good | Fair |
| Promotion – Advertising | Fair | Good | Poor |
| – Tournaments | None | None | None |
| – Leagues | None | None | None |
| – Incentives | Poor | Fair | None |
| – Appeal to Market | Good | Excellent | Fair |
| Return on Investment | Good | Very Good | Poor |

**Exhibit 2  Adult Survey Results**

- Sample Size of 100: 50 males and 50 females.
- Interviews were conducted at the Centre on Barton and with friends and colleagues.
- The respondents were informed of the miniature golf proposal and asked if they would patronize the service.
- Most frequent responses (i.e., those mentioned at least 10% of the time) were:
    - 24%   Would serve as a family activity
    - 22%   I would not use it as I don't have the time
    - 14%   Children would be more likely to come shopping with us with a miniature golf course here
    - 10%   I would play while waiting for my spouse to shop.

## Exhibit 3    Student Survey Results

Sample size was 300 students aged eleven to twenty

Gender:     Male – 144 (48%)          Female – 156 (52%)

1. Do you visit the Centre on Barton in the summer?
   Yes – 253 (84%)          No – 47 (16%)

2. a) If "Yes", would you play miniature golf?

|        | Yes      | Maybe    | No       |
|--------|----------|----------|----------|
| Male   | 99 (84%) | 7 (6%)   | 12 (10%) |
| Female | 97 (72%) | 14 (10%) | 24 (18%) |

   b) If "No", would you visit the centre for a recreational activity like miniature golf?

|        | Yes     | Maybe    | No       |
|--------|---------|----------|----------|
| Male   | 4 (16%) | 11 (42%) | 11 (42%) |
| Female | 4 (19%) | 7 (33%)  | 10 (48%) |

3. Do you think members of your family **older** than yourself would play miniature golf?
   Yes – 85 (28%)     Maybe – 154 (51%)     No – 61 (21%)

4. Do you think members of your family **younger** than yourself would play miniature golf?
   Yes – 126 (42%)     Maybe – 141 (47%)     No – 33 (11%)

5. Would you play miniature golf on a date?

|        | Yes      | Maybe    | No       | No Answer |
|--------|----------|----------|----------|-----------|
| Male   | 80 (56%) | 26 (18%) | 14 (10%) | 24 (16%)  |
| Female | 82 (53%) | 32 (21%) | 10 (6%)  | 32 (20%)  |

6. How do you view a price of $8.00 per round of miniature golf?

|        | High Price | Reasonable Price | Low Price |
|--------|------------|------------------|-----------|
| Male   | 34 (23%)   | 96 (67%)         | 14 (10%)  |
| Female | 18 (11%)   | 134 (86%)        | 4 (3%)    |

7. Would you come to the Centre on Barton on Sunday to play miniature golf?

|        | Yes      | Maybe    | No       |
|--------|----------|----------|----------|
| Male   | 70 (49%) | 32 (22%) | 42 (29%) |
| Female | 80 (51%) | 45 (29%) | 31 (20%) |

## Exhibit 4    Selected Statistics – Hamilton Census Metropolitan Area

### Population by Age Group – 2012

|  | Male | Female |
|---|---|---|
| 0 to 4 | 19,394 | 18,837 |
| 5 to 9 | 20,993 | 19,535 |
| 10 to 14 | 22,241 | 20,818 |
| 15 to 19 | 24,498 | 24,233 |
| 20 to 24 | 27,289 | 26,823 |
| 25 to 29 | 27,502 | 26,100 |
| 30 to 34 | 23,559 | 23,519 |
| 35 to 39 | 23,107 | 24,413 |
| 40 to 44 | 26,670 | 27,362 |
| 45 to 49 | 30,462 | 30,597 |
| 50 to 54 | 29,464 | 29,137 |
| 55 to 59 | 24,702 | 26,054 |
| 60 to 64 | 20,730 | 22,683 |
| 65 to 69 | 16,943 | 18,944 |
| 70 & Over | 35,043 | 49,694 |
| **Total** | **372,597** | **388,749** |

Families: Number    216,145          Households: Number    299,428
Average Size    3.0                              Average Size    2.5

### Taxation Statistics – Income Class – Individuals – 2012

| | | | |
|---|---|---|---|
| Under $5,000 | 38,220 | $50,000 to $74,999 | 86,580 |
| $5,000 to $9,999 | 36,530 | $75,000 to $99,999 | 46,550 |
| $10,000 to $14,999 | 48,650 | $100,000 to $149,999 | 25,630 |
| $15,000 to $19,999 | 47,190 | $150,000 to $199,999 | 6,800 |
| $20,000 to $24,999 | 43,170 | $200,000 to $249,999 | 2,820 |
| $25,000 to $34,999 | 67,330 | $250,000 or more | 4,220 |
| $35,000 to $49,999 | 85,290 | Total | **538,980** |

Median Income    **$33,150**

### Weather

| | Avg. # of Days With Rain | Avg. # of Days Without Rain |
|---|---|---|
| May | 11 | 20 |
| June | 10 | 20 |
| July | 9 | 22 |
| August | 9 | 22 |
| September | 10 | 20 |
| **Total** | **49** | **104** |

- Based on an accumulation of at least 0.25 millimetres. Averaged over the thirty year period 1981 to 2010. Most likely time of rainfall was 3:00 p.m. to 7:00 p.m. during these months.

## Exhibit 5    Capital Cost Estimates for Miniature Golf Course

Material Costs
- ¾" Plywood — $32.50 per sheet
- 2" x 4" Lumber — $0.62 per linear foot
- 2" x 8" Lumber — $0.99 per linear foot
- Paint — $47.00 per 3.78 litre pail
- Carpeting — $55.40 per square metre

Average Construction Cost per Hole

| Material Cost | |
|---|---|
| 125 feet of 2" x 4" Lumber | $ 77.50 |
| 60 feet of 2" x 8" Lumber | $ 59.40 |
| Four sheets of ¾" Plywood | $130.00 |
| 11.25 square metres of Carpeting Kentucky Blue Grass colour | $623.25 |
| Miscellaneous (nails, sheet metal, batteries, motors, sand, shrubbery) | $250.00 |
| Paint – one pail per hole | $ 47.00 |
| **Total Material Cost** | **$1,187.15** |

Labour Cost
- All construction to be done by the industrial arts class at the high school under the supervision of a craftsman-certified teacher — FREE

| | | |
|---|---|---|
| Total Cost per Hole | $1,187.15 | |
| Total Cost of Eighteen Holes = 18 x $1,187.15 | | $21,369 |

Other Expenses
| | |
|---|---|
| Pro Shop | $5,000 |
| Fencing | $4,450 |
| Miscellaneous (putters, balls, cards, pencils) | $3,250 |
| Total Other Expenses | $12,700 |

Total Capital Cost of Miniature Golf Course — $34,069

# LIME LIGHT CINEMA

In January 2014, after nine months of operation, Lime Light Cinema of Victoria, British Columbia, was still not generating satisfactory revenues. To attract larger audiences, Head Office in Montreal had made two major changes to the marketing strategy in the last year: it changed the programming format and the pricing strategy. Olga Siroonian, the new Manager, was faced with the responsibility for successfully implementing these changes and increasing local profitability by at least 25%.

## Company History

The theatre had operated as a pornographic film house under the name "The Cosmopolitan Cinema" for twenty-five years. In January, 2013, Phoenix Theatres of Montreal purchased the business as part of an expansion plan. To reposition the theatre as a first run art film cinema featuring two films per evening, major renovations were undertaken in late February. In March, Phoenix reopened the theatre as "The New Cosmopolitan Cinema."

From the beginning, the new cinema encountered problems with image. Even though the concept had changed, association with the previous name still branded the cinema as a place to see pornographic films. In October, the name was changed to "Lime Light Cinema" (an homage to Charlie Chaplin's silent film "The Lime Light") and a new marquee was erected. In November, the owners fired the Manager and promoted the Assistant Manager (of four months), Olga Siroonian, to the position. Prior to becoming Assistant Manager, Olga had worked for five months as one of the theatre's ushers.

## Company Problems

After the name change and under Olga's new management, business improved slightly. In March and April the average audience size had been about 60 people per show, well below the 375 seat capacity. Attendance in December was 60 to 70 people per show. Olga remarked that "As we will be receiving an average admittance fee of $5.00 per head, we will pretty well have to pack the place every night to break even."

In Olga's opinion, the theatre had two problems: 1) people did not know much about the theatre; and 2) people did not know much about the films being shown. "People don't know what they are getting when they go to see an art film," stated Olga.

After eight months of operations, Phoenix finally realized that Victoria did not have the population size or audience interest to support a "first run art film cinema." (Available cost data is outlined in Exhibit 1).

---------------------------------------

This case was written by Marvin Ryder. Case material is prepared as a basis for classroom discussion only. Copyright 2014 by Marvin Ryder, DeGroote School of Business, McMaster University, Hamilton, Ontario. This case is not to be reproduced in whole or in part by any means without the express written consent of the author.

## Recent Changes

Responding to low attendance figures, the theatre's concept was changed again. Beginning in mid-December, Lime Light Cinema became a repertory theatre featuring two different movies every night (one shown at 7:00 p.m. and the other at 9:15 p.m.). Features would include second-run commercial films (movies shown two to three months after their premiere), occasional premiere films, classic movies, foreign films, and first-run art films. Olga explained, "Most of the films shown will have been in Victoria already. People will know the films and that will make our promotion job a lot easier." The variety of films and reduced prices were key elements in the new strategy.

As well, in mid-October, Lime Light moved from a straight admission fee of $6.00 for students and $8.00 for adults to a membership basis. Company management felt the old prices were not low enough to attract enough people to come to see a film with which they were unfamiliar. Members would pay $10.00 a year to join the cinema and $4.00 per show admission fee. For non-members, the fee would be $6.00. At these prices Lime Light Cinema would be offering lower prices than the other repertory theatre in town which charged a $15.00 membership fee, and $5.00 and $7.00 for admission. Initially, Lime Light Cinema had ordered 2,500 membership cards.

There was one bright spot – in the last two weeks of December, Olga had increased theatre revenues and profits by making changes in the candy bar operation. She added/deleted products and adjusted prices on soft drinks and popcorn. The average receipt per patron had increased from about $2.50 to $3.00. The theatre generated a high contribution margin, and thus profit, from the candy bar operation, so these changes were important.

## The Theatre Industry

There were eight commercial theatres and one repertory theatre in Victoria. (See Exhibit 5 for Population data) Originally, Lime Light Cinema was not directly competing with either type of theatre. However, with the changes, it would be competing directly with the other, very well established, repertory theatre in town. Olga expected to have some initial difficulty competing for business but believed that there was room for two repertory theatres in Victoria.

Besides the other theatres, another source of competition was home DVD or DVR machines along with Netflix. The latter allowed an unlimited number of movies or television shows from an immense library to be streamed to a computer or internet-enabled television in exchange for a monthly membership fee. By the time second-run commercial films were shown at Lime Light Cinema, they were available on DVD's or online. People could purchase or rent these media for home viewing. Premiere films would not be affected by DVD sales or online services. Art films were generally not available to the public on DVD or online. Only with concerted effort could some of the exotic art or foreign film titles be located. Olga had read predictions for new technologies. Consumers had a choice between computers and tablets, iPod's and iPad's, and even smartphones as sources of entertainment. The future would likely see other new devices to compete with watching a movie in a theatre.

A variety of customers patronized the cinema. Olga estimated that 35% of her customers were students from the University of Victoria. Customers could be categorized as follows:
1) regular movie-goers who could afford to go to commercial theatres;
2) avid movie buffs, including film students from the University of Victoria;
3) people who wanted something off-beat and different; and
4) people who were just looking for an inexpensive night out.

She believed the theatre was targeting and promoting to everybody.

## Promoting the New Concept

Approximately $30,000 had been allocated to promotion for the year. In the past, most advertising had been allocated to newspapers. Lime Light Cinema placed a seven-line ad daily in the Victoria Times-Colonist and an occasional sixty-line ad in The Martlet, the university student newspaper. (See Exhibits 2 and 3) Olga felt that radio advertising was generally too expensive but would use it occasionally to promote premiere films. In addition, bi-monthly tabloid-type printed program schedules were distributed to potential customers through all record stores and donut shops in the city. These schedules were provided by Head Office (Information on media costs is provided in Exhibit 4). Lime Light had a Facebook page. It was mostly used by patrons to discuss movies. She had wondered if she should launch a website or maybe get a Twitter account to also promote the theatre. Olga did not really know if any of the advertising was effective. She did know that the future promotional strategy for the repertory concept had to be successful or she was out of a job.

The objective of the new promotional program was to make people aware of the repertory format, the new prices, and to sell memberships. So far, Olga had purchased a sixty-line ad in the Victoria Times-Colonist to announce the opening and had arranged an interview on a CHEK-TV entertainment program to talk about the new concept. She was thinking of trying to arrange a couple of radio interviews as well, but she knew more had to be done.

### Exhibit 1    Theatre Cost Information

| | |
|---|---|
| Film Rental Per Showing | $200.00 |
| Estimated Management Salaries | $45,000 per year |
| Estimated Theatre Lease (building & utilities) | $5,000 per month |

Estimated Gross Margin on 'Candy Bar' operation - 65%
(i.e., 65 cents of each dollar spent at the 'Candy Bar' was profit; 35% covered variable costs)

Average Staffing Per Night (average of 3.5 hours per person at minimum wage)
    One Cashier
    One Candy Bar Person (Two on Friday and Saturday nights)
    Two Ushers
    One Doorman
    One Projectionist – 4.5 hours per night at $20.00 per hour

**Exhibit 2**  **Daily Ad in the Victoria Times-Colonist**

**Exhibit 3**  **Weekly Ad in The Martlet (University of Victoria Newspaper)**

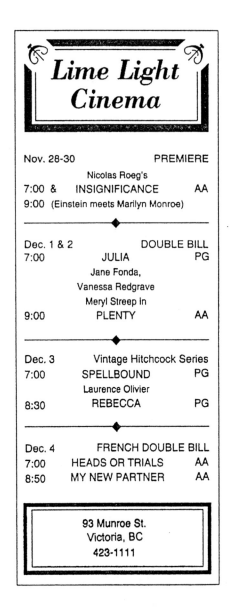

## Exhibit 4   Advertising Rates

| NEWSPAPER | Circulation | Line Rate (per column) |
|---|---|---|
| Victoria Times-Colonist | 88,796 (total paid daily) | $5.16<br>$4.61 (over 10,000 lines per year) |
| TV Plus (Victoria edition)<br>(free distribution) | 42,913 (weekly)<br>(26 week schedule) | 1/4 page - $310.00/week |
| The Martlet | 10,000 (weekly) | $2.62<br>$2.48 (rate for weekly contract) |

| RADIO | AAA | AA | A | B | |
|---|---|---|---|---|---|
| CFAX (#1 station in Victoria | $55 | $43 | $35 | $28 | 60 Seconds |
| area - reach 28% of homes) | 45 | 34 | 29 | 23 | 30 Seconds |

AAA - 6:00 am to 10:00 am weekdays           AA - 4:00 pm to 11:00 pm weekdays
A - 10:00 am to 4:00 pm weekdays and Saturdays    B - all other times

    Reach plan    $35 per spot for 21 sixty-second spots
                    Made up of four AAA, six AA, six A and five B spots

CFUV University of Victoria - 60 Seconds - $10.00

## Exhibit 5   Population Statistics

| Metropolitan Victoria | | |
|---|---|---|
|     Population | | 344,615* |
|     Rank in Twenty Largest Canadian Census Metropolitan Areas | | 15th |
|     Age Groups | | |
|         14 and under | 45,190 | |
|         15 to 24 | 42,535 | |
|         25 to 34 | 45,200 | |
|         35 to 44 | 42,825 | |
|         Over 44 | 168,865 | |
|     Average Household Income per week | | $1,525.96 |
|     Rank in Twenty Largest Canadian Census Metropolitan Areas | | 9th |

\* Student population
    University of Victoria  19,500

Source:  Statistics Canada, Canadian Census, 2011

# Wil's Grill

*Leonard R. Hostetter, Northern Arizona University*
*Nita Paden, Northern Arizona University*

In January 2017, John Christ needed to make some decisions about his business, Wil's Grill. Not long ago, his dad had said, "Son, passion has gotten you here; not the money." Now, John needed to focus on "the money" – but which path should he take? He could expand his "street food" business, add a catering business, or do something else. John, who loved to make customers happy by serving them great healthy local food, recognized that he also needed to do so profitably.

## BACKGROUND

John grew up on a ranch in Cave Creek, AZ, a small community northeast of Phoenix Arizona. His parents had food service and restaurant experience, and cooking and entertaining were an integral part of spending time with them. "By age 10," John recalled, "I could cook."

As a teenager, John bussed tables at a restaurant where his dad Wil worked. He also spent many mornings with his dad at a clay-bird sport shooting range near Cave Creek. When done, they needed to go elsewhere for lunch, since the range did not offer food or beverages. So, father and son worked out an agreement with the range owner to open a small food booth on-site, which they named "Wil's Grill." On a single grill they cooked burgers, fries and served beverages. Wil taught his son the nuts and bolts of running the business: obtaining necessary permits and licenses, ordering food and supplies, shopping, transportation, inventorying, cooking, cleaning and most importantly, "treating customers as friends." Hospitality-driven service was a core value.

To celebrate his high school graduation in December 2009, John went on a 30-day backpacking excursion with the National Outdoor Leadership School in Wyoming, where he later recalled, "I honed my leadership skills there and this would serve me well in managing my future business."

In August 2010 Wil closed Wil's Grill when John enrolled at Northern Arizona University (NAU), in Flagstaff, about 120 miles north of Phoenix. At that time NAU

---

Copyright © 2017 by the *Case Research Journal* and by Leonard R. Hostetter, Jr. and Nita Paden. This case study was prepared as the basis for classroom discussion rather than to illustrate either effective or ineffective handling of an administrative situation. The authors wish to thank John Lawrence, Brent Beal, Gina Grandy, Janis Gogan, Kathryn Savage, Lance Rohs, Joseph Anderson and the anonymous CRJ reviewers for their helpful suggestions on how to make this a more effective case. An earlier version of the case was presented at the 2016 Annual Meeting of the North American Case Research Association in Las Vegas, NV, United States.

enrolled about 23,000 students. John majored in Environmental Studies, and also took classes in other areas, driven by "my inquisitive nature to learn as much as I could about the world around me." At the NAU School of Hotel and Restaurant Management John learned about the "clean food" movement – characterized by locally produced, organic foods and sustainable practices.[1] Clean food was healthy for both the planet and for people through production of efficient amounts of food, provision of leftovers to local shelters, and minimization of waste via biodegradable products and recycling practices.

## WIL'S GRILL FLAGSTAFF

On a visit to Costa Rica in 2013, John and another NAU student – Karl Shilhanek observed a vibrant "street food" community.[2] The "chicken lady," "kabob guy," and many other vendors served tasty, locally sourced and ready-to-eat fresh foods to local residents on the street, in the market, at a fair or other public place. Vendors sold "street food" from a portable stall, cart or food truck. John and Karl were inspired to start their own business, and the flames of Wil's Grill were reignited when they founded their own Wil's Grill in Flagstaff, AZ in January 2014.

The young men worked hard to get Wil's Grill off the ground. They wrote a business plan, secured the required permits and licenses, and set up as a general partnership. The two partners each invested $500 to get the business off the ground, and John's parents provided a $2000 low-interest loan to help them purchase grilling equipment.

"We earned our stripes in the first year," John recalled. "We were hands-on with every aspect of the business." Karl focused on business strategy, marketing and social media. He created a website that included their "clean food" menu, a mobile app and a social media presence (on Facebook). John focused on operations and food preparation. He established relationships with five local food sources -- including John's parents' Happy Mountain Farms. John believed his relationship with farms and producers "allowed me to have a unique understanding of the local supply chain."

Wil's Grill was highly portable, and targeted two main markets: 1) NAU students who were tired of chain-based fast food and wanted good, reasonably priced, late night food, and 2) community events, where organizers and customers wanted reasonably priced, clean, high quality street food (in contrast, many street food vendors served manufacturer prepared and processed food). Operations included procuring food, preparing main courses and sides, transporting food to venues, and hiring temporary labor for serving and clean-up. Wil's Grill leased excess kitchen space in non-competing Flagstaff restaurants and bars, for prepping or cooking some food. Once the food was prepared in these locations it was served on tables with warming trays. For outdoor events, an event management company assigned Wil's Grill and other vendors to specific locations for specific hours. Most food (e.g. burgers, vegetables) was prepared on site, in view of customers.

Within four months John and Karl were able to pay off the $2000 loan; since then, they had taken no further loans. The business was not profitable and they did not pay themselves a salary. John and Karl both worked second jobs to cover basic living expenses in 2014 and 2015, and their parents paid their college tuition. John lived a simple lifestyle with minimal financial obligations. They did not invest in a brick-and-mortar operation. Their "office" was as portable as the business.

In May 2014, Karl decided to relocate to Bellingham, WA to be closer to his family. The breakup was amicable. John reestablished Wil's Grill as a sole proprietorship. Without his partner, at first John relied on "gut instinct" to run his business. Summer 2014 was tough, especially interviewing and hiring people. John felt this "was challenging. I didn't know what I was looking for." To hire temporary employees for street events he posted ads and networked with local bar owners. In June John hired what he referred to as "my first permanent part-time employee, Cody McCrae, a Hotel and Restaurant Management student." Cody had also "grown up in the kitchen." On his first day John gave him some instructions and left for another commitment. Working alone, Cody prepared sliders and coleslaw and proved himself. John placed a lot of trust in Cody, his first assistant manager. Cody flexed his hours and worked as business levels demanded.

Preparation and cooking was fast paced, whether in a leased kitchen or on the grill at an event. There were many 18 to 20 hour work days. John believed that he treated his temporary employees fairly, and therefore they were customer focused and wanted to work for him again. John also learned that he needed to define routines and flow-chart responsibilities for some job positions, and to calculate staffing based on the estimated number of plates/day to be served.

## STREET FOOD EVENTS

Street events involved lots of guess work, since both weather and attendance were unpredictable. John told a friend, "It's like rolling the dice to try and guess what food people will want." During the Flagstaff Pro Rodeo, Wil's Grill served 425 plates of barbecue per day, whereas for most events, 200 - 300 plates/day was typical. During the Rodeo, one employee quit. John recalled: "Lines formed quickly; everyone came to the booth around lunch time, hungry. We performed well -- though there's always room for improvement." Getting food out quickly was most important, and food quality was more important than presentation or quantity.

Customers enjoyed watching food preparation, including the employee chatter. Pricing was customized for each event client (Exhibit 1 includes sample menus), so event revenue varied. A 200 to 300 plate day could gross $2,000 to $3,000, enough to sustain operations. Ongoing grill maintenance and food purchases were the main operating costs. Other expenses included liability insurance, permitting, licensing and payroll. John lived modestly, paid bills in cash and avoided debt. He used his personal pick-up truck to transport food, and budgeted for fixed costs, irrespective of ebbs and flows of revenue. He estimated that profit margins averaged 18% - 25% -- good for the street food business. 2014 to 2016 revenues totaled $129,000 (Exhibit 2).

John had "learned on the fly;" he worked hard and wasn't discouraged by challenges. Feelings he experienced when customers told him how much they enjoyed his street food and his passion for clean food outweighed any discouragement. Street food was fun and fast-paced. John loved it.

## THE WIL'S GRILL MARKET

By 2015 Wil's Grill primarily served Flagstaff, along with Prescott and Sedona to the south, Williams to the west and most of Northern Arizona (with a combined population of about 275,000 people[3]). Winter weather limited the number of street food events in Flagstaff, given its 7,000 foot elevation. Sedona, Prescott and the Verde Valley (all within 100 miles of Flagstaff), at lower elevations, were warmer. Collectively, each of these communities held almost 50 events that featured street food. For special events John sometimes traveled as far as Phoenix or Page (both within 150 miles of Flagstaff); he included fuel costs in his pricing. Phoenix was the 12th largest metropolitan area in the U.S. with a population of 4.57 million people and a vibrant street food scene.[4] Wil's Grill had received excellent reviews from local writers, food critics and customers, and was featured in the July 2015 issue of *Flagstaff Business News*. More food trucks were also appearing on the scene, and some new entrants served healthier fare. John's promotional marketing budget for 2016 was $2,100, although he believed he should spend $5,000.

As for catering, large competitors included Big Foot BBQ and Satchmo's (local barbeque restaurants that also offered catering), as well as Main Street Catering and Thorneger's Catering. Some competitors had been in business for twenty years or more, and were well-established in the local catering market. Wil's Grill had the strongest focus on clean food, and John received referrals from caterers for specialties Wil's Grill was known for -- smoked meats and barbeque.

Various studies conducted in the United States indicated a growing interest in "clean food" and this was beginning to influence some customers' food and beverage purchase decisions. Consideration for healthy choices had reportedly increased from 61% in 2012 to 71% in 2014, and in 2015 67% of respondents had given thought to environmental sustainability, 72% had given consideration to how food was produced or farmed, and 26% regularly purchased locally sourced items.[5] John believed that the demographics and psychographics of people in Northern Arizona aligned well with the national clean food movement.

## THE CATERING MARKET SEGMENT

In fall 2015, John coordinated with NAU marketing research students on an exploratory survey to learn more about customer perceptions of the Wil's Grill brand, food offerings, the clean food movement and the catering market segment. He realized that with just 79 respondents, the survey results were directional at best.[6] The survey indicated that 56% of respondents were willing to spend at least $11/person on a catered event, of which, 24% were willing to spend about $16/person and 78% were willing to spend an additional $1 to $6/person for clean food. 72% of respondents had never heard of Wil's Grill. Regarding the decision to use a caterer, barbeque beef and pork, Mexican, Italian, Asian and vegetarian were the most desired catered event food options. A caterer's reputation, customer reviews, service, food selection and price were critical factors in selecting a caterer, and 63% indicated that locally sourced food would influence their decision.

To keep current on trends and opportunities, John was a voracious reader of food trade journals. One article stated that "farm to fork has been a trend emerging in weddings."[7] Catered events could have margins of up to 40%. Catering customers were typically older, more affluent, and included both individuals and businesses. The business model was somewhat more predictable than street food vending, with a pre-determined number of guests, food type, pricing, and event specifics. Catering opportunities were available year-round in the Northern Arizona market.

Catering could be labor intensive. John estimated that a buffet-styled catered event required one staff member for every 30 guests, and a "plated, waited and served" event would need up to twice as many staff members. New job descriptions and training would need to be developed. John expected that he would need to expand his menu and improve the food presentation, based on what clients wanted. Customers also often asked caterers to provide décor, entertainment, etc.

John would need to invest in new kitchen equipment, logistics, and a cargo trailer to store, maintain and transport food. The required investment would increase if John planned to cater multiple events simultaneously. John realized he'd need to get out of his comfort zone and assume debt to expand into the catering segment. He would have to figure out a way to secure financing.

John estimated his annual catering marketing expenses would be $7,500. He believed he would realize synergies with his existing street food segment marketing investment, but a catering client base would need to be developed. John saw his brand as "Wil's Grill and not Wil's Barbeque, offering a wide assortment and variety of foods and flavors off the grill. We can be so much more than barbeque." In street food, John focused on smoked meats and barbeque because it was easily prepared and sold at a reasonable price. Wil's Grill stood for street food among those who were aware of the brand. His motto "Have grill will travel" reflected that he traveled to various street food events. He had never copyrighted this motto, but he had trademarked the Wil's Grill name.

John developed high-level ballpark estimates for future cash flows and investment associated with the options under consideration for growing the business (Exhibit 3). He strongly believed that the Wil's Grill brand was defined by "our reputation among those we've served, and those who have heard about us. Our reputation is one of sincerity, transparency, consistency and quality."

John needed to make a strategic decision: how to move forward with Wil's Grill and his livelihood?

# NOTES

[1] Feine, Suzy. *Green, The New Color of Love*. CaterSource, 1 May 2009. Web. 24 July 2017. http://www.catersource.com/green-catering/green-new-color-love.

[2] "What Is Street Food?" *The Street Food Institute*. N.p., n.d. Web. 24 July 2017. http://www.streetfoodinstitute.org/what-is-street-food/.

[3] Arizona Cities by Population. (2015, May). Retrieved July 24, 2017, from https://www.arizona-demographics.com/cities_by_population; United States Census Bureau/American Fact Finder.

[4] Theobald, B. (2015, March 26). Census: Phoenix area population grew rapidly. Retrieved July 24, 2017, from http://www.azcentral.com/story/news/arizona/politics/2015/03/26/census-phoenix-area-population-grew-rapidly/70507534/.

[5] http://ljournal.ru/wp-content/uploads/2017/03/a-2017-023.pdf. (2016). *International Food Information Council Federation - Food and Health Survey 2015*, (10), 69-77. doi:10.18411/a-2017-023.

[6] Survey Monkey September 2015 – *Wil's Grill* Case Author and MKT 439 Marketing Research Students; 79 respondents to a 20-question survey.

[7] Jacobs, A. S. (2016, March 29). Relaxed Luxury: New Farm-to-Fab Wedding Inspiration. Retrieved July 24, 2017, from http://www.instyle.com/news/relaxed-luxury-new-farm-fab-wedding-inspiration.

**Exhibit 1 – Wil's Grill Sample Menu Items**

| | | | |
|---|---|---|---|
| Marinated Chicken & Veggie Kabob | $5.00 | Pork & Brisket Sandwich | $12.00* |
| Grass-fed Hamburgers | $10.00 | Grilled Chicken Legs | $5.00 |
| Beer Brat with Sauerkraut | $8.00* | Loaded French Fries | $6.00* |
| Mac & Cheese, Cole Slaw, Beans | $3.00* | Gatorade | $2.00 |
| Bottled Water | $1.00 | | |

*Price varied based upon the vendor fee charged by event management.
Note: John targeted a minimum avg. ticket order of $10.00 and a 25%-35% food cost.
Source: John Christ – Owner, Wil's Grill (July 2017) – *Sample Street Food Menu*

Example – BBQ Lunch Garden Party for 30 people; Buffet Service *Sample Catering Menu*

| | Quantity | Unit Price | Line Total |
|---|---|---|---|
| Appetizer – Garden Fresh Bruschetta | 1 tray | $120/tray | $120.00 |
| Salad – Mixed Field Greens | 30 cnt | 3.50/cnt | $105.00 |
| Entrée – Slow Smoked Brisket | 8 lbs | 22.99/lb | $183.92 |
| Entrée – Slow Smoked Turkey | 8 lbs | 22.99/lb | $183.92 |
| Side – Buttermilk Cornbread | 3 trays | 4.99/dzn | $37.50 |
| Side – Mama's Tater Salad | 7.5 lbs | 9.99/lb | $74.93 |
| Side – Cowboy Beans | 7.5 lbs | 7.50/lb | $56.25 |
| BBQ Sauce | 0.75 gal | 20.00/gal | $15.00 |
| Beverage – Unsweetened Tea | 1.5 gal | 5.50/gal | $8.25 |
| Beverage – Fresh Squeezed Lemonade | 1.5 gal | 10.00/gal | $15.00 |
| Services – On-site Buffet | 1 hour | 125.00/hr | $125.00 |
| | | Sub-Total | $924.77 |
| | | Tax at 10.95% | $101.26 |
| | | Grand Total | $1,026.33 |

Notes: Menus available on Wil's Grill website.
    Clean food discussed on website and menu boards at events.
Source: John Christ – Owner, Wil's Grill (April 2017)

## Exhibit 2 – Profit and Loss Statement (2014-2016)

| Ordinary Income/Expense ($ US) – Accrual Basis January through December | | | | |
|---|---|---|---|---|
| | 2014 | 2015 | 2016 | Comments |
| **Income** | | | | |
| Food Sales | 18,000 | 24,568 | 86,921 | |
| Total Income | 18,000 | 24,568 | 86,921 | 2014: 8 special public events & seasonal weekend street service; 2015: 15 special public events; 2016: 40 special public events |
| **Cost of Goods Sold** | | | | |
| Food Purchases | | 12,533 | 32,401 | Goal: = 30% of Food Sales thru purchasing efficiencies and sourcing co-operative |
| Direct Labor Payroll (Cody paid $3,613 (2015) and $6,470 (2016)) | | 4,937 | 10,675 | 2015: Cody (450 hrs) & sub-contract labor<br>2016: Cody (650 hrs) & sub-contract labor |
| Business Licenses/Permits/Insurance | | 1,434 | 4,005 | Per event basis |
| Total COGS | 6,000 | 18,904 | 47,081 | |
| **Gross Profit** | 12,000 | 5,664 | 39,840 | |
| **Operating Expense (Fixed)** | | | | |
| Advertising & Promotion | | 1,105 | 2,131 | |
| Automobile Expenses | | 369 | 4,700 | Fuel and maintenance |
| Bank Service Charges | | 287 | 210 | |
| Computer and Internet Expenses | | 228 | 1,000 | |
| Office Supplies | | 283 | 53 | |
| Professional Fees | | 2,043 | 2,300 | Client meetings, Legal, Acct, R&D |
| Propane | | 286 | 555 | |
| Reimbursement | | | 109 | |
| Rent Expense | | 1,412 | 1,500 | Leased kitchen space |
| Repairs and Maintenance | | | 92 | |
| Restaurant Supplies | | 1,454 | 549 | |
| Supplies | | 382 | 1,239 | |
| Uniforms | | | 73 | |
| Utilities | | | 255 | |
| Total Operating Expense | 15,000 | 7,849 | 14,766 | 2014 Operating Expense not itemized |
| Net Ordinary Income | | (2,185) | 25,074 | |
| **Other Income/Expense** | | | | |
| Ask My Accountant | | 33 | 1,604 | |
| Total Other Expense | | 33 | 1,604 | |
| **Net Operating Income** | (3,000) | (2,218) | 23,470 | John paid himself a salary from Net Income after reinvesting back in business (2016) |

Source: John Christ – Owner, Wil's Grill (July 2017)

**Exhibit 3 – Estimated Investment Levels and Future Revenue Projections (rounded)**

| 2016 Actual Revenue | Est. Investment | 2017 Total/ YOY Rev. | 2018 Total/ YOY Rev. | 2019 Total/ YOY Rev. |
|---|---|---|---|---|
| (A) Expand Street Food: $87,000 | $9,000 | $122,000/$35,000 | $162,000/$40,000 | $212,000/$50,000 |
| (B) Add Catering and Maintain Street Food: $87,000 | $25,000 | $147,000/$60,000 | $217,000/$70,000 | $307,000/$90,000 |

John research assumed:
- 20%-25% (A) & 40%-45% (B) profit before taxes
- discount rate range 5%-18%
- 20% YOY "normal" growth rate for street food revenue (2016-2019)
- +2% YOY inflation rate; +3% YOY contingency expense
- $3,500 revenue per catering event (2017)
- 5% YOY higher operating expenses for (B) vs. (A)
- 2017 COGS 50% (A) and 52% (B) of Total Income bef. inflation/contingency
- "Est. Investment" primarily kitchen equipment

Source: John Christ – Owner, Wil's Grill (July 2017)

case twenty-two

# "Greener Pastures": The Launch of StaGreen™ by HydroCan

*Anne T. Hale*

Stone Age Marketing Consultants was founded five years ago by Cari Clarkstone, Karen Jonestone, and Robert Sommerstone. Their target clients were small, startup firms as well as medium-sized firms looking to expand operations. Their newest client, HydroCan, had a meeting scheduled for the following afternoon and the three founders were discussing the results of their market analysis. HydroCan was a startup company that was obtaining patents in both the United States and Canada for a new type of lawn-care product. Since the company was made up of four agricultural engineers and a financial accountant, they were in need of marketing advice concerning their new product, StaGreen™. This product, when applied to most types of grass, enabled the root system to retain water longer, thus reducing the need for both extra watering and frequent fertilizing. They were anxious to take this product to market; however, they desperately needed answers to several questions, including which segment to target, how to position their new product, and what type of launch strategy they should use. They approached Stone Age Marketing Consultants approximately four weeks ago with their needs. The marketing consultants had analyzed the markets, costs, prices, and communications options. Their last task was to formulate a comprehensive strategy for the launch of StaGreen.

---

Prepared by Anne T. Hale, formerly Visiting Assistant Professor of Marketing, Faculty of Business, University of Victoria, Victoria, British Columbia, as a basis for class discussion. Copyright © 1996 by Anne T. Hale. Reprinted with permission.

## INITIAL MEETING WITH HYDROCAN

During the initial meeting between HydroCan and Stone Age, the engineers outlined the product and its potential benefits. The product was very similar in appearance to most brands of common lawn fertilizer. In fact, StaGreen was classified as a chemical fertilizer, but with one very important difference. Its primary benefit was its effect on the root system of most of the common types of grasses used for lawns. The small pellets attached to roots and attracted and retained moisture. Extensive laboratory testing demonstrated that StaGreen reduced the need for manual watering on most types of grass by up to 40%. Obviously such a product would have high demand. The first question that HydroCan needed addressed was what market to target initially with this product. Gary Gillis, CEO of HydroCan, wanted to target the consumer lawn and garden market as their initial target segment. Carla Humphreys, on the other hand, was more inclined to target the commercial lawn and garden market. Since these two markets required very different launch strategies, selecting the appropriate segment was the primary concern. And, due to the fact that both Mr. Gillis and Ms. Humphreys were extremely biased toward their position, the consultants knew that they would have to present strong reasons to support their recommendation. To make this task manageable, they divided the research and analysis along the following lines: Cari Clarkstone was to investigate the viability of a consumer launch, Karen Jonestone was to investigate the viability of a commercial launch, and Robert Sommerstone was to obtain all necessary financial information.

## THE CONSUMER MARKET

In 1995, Canadians spent nearly $2.3 billion, at the retail level, on gardening. This figure includes $945 million for grass (both sod and seed), trees, and plants, $620 million on lawn maintenance (fertilizers accounting for 52% of the total), and $815 million on hand tools, pots, window boxes, books, magazines, landscaping services, etc. In other words, gardening is big business in Canada. Lawn care is, however, a highly seasonal business, with 70% of sales occurring in the second and third fiscal quarters (i.e., April to September).

According to Cari Clarkstone's research, if HydroCan was to target this segment, they would be competing primarily with fertilizers. The consumer fertilizer market is extremely competitive, with the top two firms, Scotts Co. and Ortho Chemicals, controlling approximately 50% of the total consumer market. Both firms are headquartered in the United States (with divisional offices in Canada), and both have extensive international operations. The market share leader is Scotts Co., with their two powerful brands, Turf Builder® and Miracle-Gro® (acquired in May of 1995 from the privately held Stern's Group). Turf Builder is a slow-release fertilizer that reduces the number of applications required for a healthy lawn. This slow-release technology is relatively new—having been available to the consumer market for less than two years. Slow-release simply means that the fertilizing chemicals are released gradually over a number of months. Thus one application of slow-release fertilizer could last for a maximum of two years (although most manufacturers recommend applications every year).

Turf Builder is priced slightly lower than most Miracle-Gro products, which are advertised as maximum-growth products, and not specifically (i.e., exclusively) aimed at lawn care. Ortho's products are priced competitively with Turf Builder—their added value comes from the inclusion of pesticides within the fertilizer that prevents most common lawn infestations. See Exhibit 1 for pricing information on the major branded fertilizer products.

| EXHIBIT 1 | Competitor Prices for the Consumer Market | |
|---|---|---|
| | Size(s) | Retail Prices(s) |
| Scotts Turf Builder | 10 kg | $24.50 |
| Scotts Turf Builder | 25 kg | $59.99 |
| Scotts Turf Builder | 5 kg | $14.75 |
| Miracle-Gro–plant/crystals | 200 g | $ 8.50 |
| Miracle-Gro–lawn/garden | 2.5 kg | $12.95 |
| Miracle-Gro–liquid | 1 L | $ 7.99 |
| Ortho (with pesticide) | 10 kg | $23.99 |
| Ortho (with pesticide) | 30 kg | $68.79 |

Market research has shown that four out of ten consumers in this market have no concrete brand preferences. They rely heavily on in-store advertisements and sales staff for information and recommendations. Many consumers cannot recall a brand name or a manufacturer of fertilizer. The product with the highest brand-name awareness is Miracle-Gro; however, most associate this brand name with their plant foods rather than their lawn fertilizers. Because of consumer behaviour and attitude toward this product category, most manufacturers relied on a strong push strategy.

Most lawn care products are sold by three distinct types of retailers: discount stores, such as Canadian Tire, Wal-Mart, and Sears; specialty stores, including nurseries; and home improvement stores. The discount stores, who buy direct from manufacturers, place strict requirements on their orders and expect price concessions and special support. Marketing expenses for both Scotts and Ortho went up by approximately 10% between 1994 and 1995, with the bulk of the increase devoted to promotions to discount retailers. This indicates the relative importance of this channel—it is estimated that 60% of all consumer fertilizer sales are made in discount stores, compared to approximately 30% of sales being made in specialty stores and 10% of sales being made in home improvement stores. Discount stores have, in fact, been spending millions in renovations in order to accommodate larger lawn and garden areas within their stores. The same is true with home improvement stores, such as Home Depot, which has 21 locations in Canada.[1]

Specialty stores, the vast majority of which are nurseries, tend to be independently owned and thus much more numerous. While the 9 top discount chains across Canada control over 89% of all sales from discount stores, the top 50 specialty garden stores account for less than 28% of all sales from this store type. The most recent research indicates that there are over 1000 specialty garden stores in Canada. Most of these stores purchase from large horticulture wholesalers, and receive little, if any promotional assistance from the major manufacturers. Home improvement stores are growing in numbers, and tend to be large, powerful chains, such as Home Depot. While these stores do not represent a large portion of current sales, they are expected to grow in importance. Like discount stores, home improvement stores buy direct from the manufacturers and require price concessions and promotional support.

The large manufacturers of fertilizer products generally spend approximately 20% of sales on marketing activities. The bulk of this money goes toward the sales force, selling in general, and trade promotions. Due to the three different channels in which their product is sold, most fertilizer manufacturers recognize the importance of a strong sales force. In terms of trade promotions, they provide in-store literature, displays, and sales training—especially to the large discount stores and the home improvement stores. Less important is advertising. Miracle-Gro is the most heavily advertised brand on the market, and Scotts generally spends 4% of sales on advertising (which probably accounts for the high brand-name awareness). Scotts advertises TurfBuilder, but only during the early spring when demand for lawn fertilizers is at its peak. Most companies run their advertisements for their existing brands and any new brands they may be launching during the spring and early summer months. Thus, advertising expenditures are generally at their highest in March, April, May, and June, and zero at all other times. Only Miracle-Gro is advertised year-round, with different messages at different times of the year. For example, Miracle-Gro advertises its benefits for house plants during the winter months, and its benefits for fruits, vegetables, and flowers during the spring and summer months.

## THE COMMERCIAL MARKET

The commercial market consists primarily of Canada's 1800 golf courses, but also includes commercial properties such as office complexes and apartment buildings. The most lucrative market, however, are golf courses. Currently under fire for being a major source of groundwater pollution, due to the high and frequent levels of fertilizers used to keep courses green, most owners are actively looking for ways to cut both water and fertilizer usage. Course owners spend, on average, $300 000 to maintain their golf course during the year, of which 42% represents water usage costs and 24% represents fertilizer purchases. For extremely large, complex courses, this figure can run as high as $800 000, and for smaller inner-city public courses, as low as $104 000. Tests have indicated that StaGreen™ will reduce water usage by one-half and fertilizer usage by one-third. This is the primary reason why Ms. Humphreys was so adamant that the company select the commercial market as its primary target.

The game of golf has been enjoying a renewed popularity after a drastic decrease in participation during the 1980s. The growing number of public courses with reasonable fees, the continued aging of the Canadian population, and the development of better equipment have all contributed to this growth in popularity. It is estimated that the number of golf courses will increase by 22% to 2200 within five years. Most golf courses are independently owned and operated. Only 4% of all courses are owned by a company that owns more than one course. Courses are dispersed throughout Canada, but British Columbia, and Vancouver in particular, boast the highest number of courses.

Currently, golf courses purchase maintenance supplies from wholesalers who specialize in products uniquely designed for the type of grasses used. Manufacturers of these fertilizers tend to be small firms, or divisions of the larger chemical companies. The market share leader in golf course fertilizers is Sierra Horticultural Products, a subsidiary of Scotts Co. Scotts purchased Sierra in 1993, and it represents only about 2.2% of Scotts' total sales. Their biggest competitor in Canada is Nu-Gro Corporation, an Ontario-based horticultural products company founded in 1992. Unlike the firms competing in the consumer lawn maintenance market, these firms spend only about 9% of sales on marketing activities. These firms engage in little advertising, preferring to spend their marketing funds on sales calls to golf courses. They provide free samples of their products to non-users and try to build solid, long-lasting relationships with course owners. They know that it takes a tremendous selling effort to get a golf course owner to switch brands. If satisfied with their current brand, many course owners are unwilling to risk switching to a new product that may not perform as well. Since the condition of the course is the most important attribute in a consumer's selection of a course to play, course owners tend to be highly brand-loyal.

Course owners, however, have two overriding concerns. The first concerns the growing public debate on the groundwater pollution caused by golf courses. Heavy use of fertilizers and constant manual watering results in a chemical buildup in nearby reservoirs. In fact, according to the U.S. Environmental Protection Agency, golf courses are the major source of groundwater pollution in the United States. More and more negative publicity, in the form of newspaper and magazine articles, has resulted in golf course developers being denied permits to construct new courses. Thus, addressing the issue of groundwater pollution is a major concern with course owners.

Their second problem is that of shrinking profits. While golf is growing in popularity, and more courses are being built to accommodate demand, the actual number of golfers that can be accommodated on any one course cannot be expanded. With some courses engaging in green-fee price wars, profit margins for many of the public courses have become strained. Thus, while loyalty may play a role in fertilizer purchases, these difficult problems will also influence purchase behaviour.

Estimated to be about one-eighth the size of the golf course market is the balance of the commercial lawn care market, consisting of apartment and office complexes. Their needs are much less complex than those of golf courses, resulting in purchasing behaviour that mirrors that of the consumer market. Little concrete information is available concerning the number of office complexes and apartment buildings, although estimates have put the total figure around 2900, of which 16.5% represent multiple holdings by one corporation. These commercial real estate property firms spend a disproportionate amount on lawn maintenance—they account for nearly 26% of the total dollars spent in this sector of the commercial maintenance market. This sector of the commercial market tends to purchase in bulk through wholesalers—generally the same wholesalers who service the specialty stores in the consumer market.

## HYDROCAN

HydroCan was incorporated nearly one year ago. They have leased their production facilities, and have purchased and/or leased all of the equipment and machinery necessary for use in the production of StaGreen. Their production facility has the capacity to produce 180 000 kilograms of StaGreen per month. The owners of HydroCan have suggested a quality/value-added pricing strategy. They believe that they have a superior product that will save the end user both time and money, due to the reduced need for fertilizer products and manual watering. The founders of HydroCan outlined their ideas for the launch year marketing strategy for both the consumer and the commercial lawn-care markets.

If HydroCan elects to target the consumer market, they will package StaGreen in a 10-kilogram bag, which market research indicates was the most popular size with consumers. They will set their price to trade (i.e., wholesalers and retailers) at $22.50, with their variable costs representing 52% of sales. On average the large discount stores and home improvement stores take a 25% markup on lawn maintenance products. The smaller specialty stores take a larger markup of 35%. Wholesalers (if used) take a 15% markup. Fixed production costs include $700 000 in annual rental (for the site and equipment), general and administrative expenses of $80 200, research and development expenses of $20 650, and miscellaneous expenses of $12 350. Distribution costs (including freight, warehousing, and storage) represented a significant yearly expense due to the seasonal nature of demand. Production of StaGreen would be continuous year-round; however, sales would be highly concentrated in the months of April through September. This means that the company would have relatively high distribution costs, estimated to be $426 000 per year. Not yet included in any of their financial statements are the salaries for the four founding partners of HydroCan. They would like to earn $50 000 per year (each), but are willing to forgo their salaries in the launch year.

Their marketing budget has been set at $555 000, and HydroCan has suggested this amount be allocated to the various tasks, as shown in Exhibit 2. Seasonal discounts are price discounts offered to retailers and wholesalers as an incentive to purchase well in advance of the peak selling season. HydroCan plans to offer these discounts, estimated to be 20% off the trade price for each bag purchased, to wholesalers and retailers in the months of November and December as a method to reduce warehouse and storage costs. The displays will cost approximately $250 each (which includes promotional materials, such as brochures), and will be furnished to discount stores, home improvement stores, and as many nurseries as possible. The sweepstakes is used to increase awareness and interest in StaGreen. Consumers will have the chance to win several valuable prizes including a year of free lawn maintenance, lawn and garden equipment and supplies, and other related prizes.

| EXHIBIT 2 | Allocation of Marketing Budget for Consumer Market Launch |
|---|---|
| Marketing Task | Total Expenditure (estimates) |
| Seasonal discounts | $225 000 |
| In-store displays | $ 92 000 |
| Magazine advertising | $104 000 |
| Newspaper advertising | $ 84 000 |
| Sweepstakes | $ 50 000 |

In terms of the sales force, HydroCan has planned on hiring 20 sales reps at an average cost of $25 000 per rep (salary and commission). The sales reps will be responsible for selling the product to the various channels as well as offering sales training seminars.

If HydroCan elects to target the commercial market, then the size of the product will be increased to a 50-kilogram bag, which they will sell to wholesalers or end users at a price of $150. Because they would be charging a slightly higher price under this option, variable costs as a percentage of sales drop to 40%, resulting in a relatively high contribution margin of 60% of sales. Wholesalers, who generally sell directly to the commercial users, take a 15% markup. Fixed expenses will remain nearly the same as for the consumer market option, with the exception of marketing and distribution costs. None of the promotional activities, such as displays, seasonal discounts, sweepstakes, or advertising will be used in the commercial market. Instead, the size of the sales force will be increased to 30 to handle the lengthy sales calls necessary to golf courses. In addition, $100 000 has been set aside for free samples to be distributed to potential customers by the sales force. Finally, distribution costs decrease if the commercial market is chosen because demand tends to be slightly less seasonal. Thus costs for freight, warehousing, and storage decrease to $225 000 under this option.

## THE DECISION

The three founding partners of Stone Age Marketing Consultants were in the conference room discussing the results of their research and analysis. As Karen Jonestone pointed out, "A strong case can be made for both target markets! Each has its own advantages and limitations." Rob Sommerstone countered with the fact that HydroCan was a startup business. "Their financial resources are extremely limited right now. They cannot increase their production capacity for at least two years, and if they hope to acquire expansion capital to increase their total capacity, they need to show a profit as early as possible." Cari Clarkstone was considering a more creative solution—targeting selected parts of either or both the consumer and commercial markets. Before the group could begin to assess the viability of HydroCan and its product, StaGreen, they had to decide on which market to target and how to position StaGreen in that market, and then they had to develop a viable marketing strategy for the launch year. The final pressure for the group was the fact that HydroCan needed to launch in February—just prior to the peak selling season; thus, the consultants knew there was no time to acquire additional market research. The decision had to be based on the information at hand.

## Note

1. *Maclean's,* April 22, 1996, pp. 62–63.

# BRIEF CASES

4240
JULY 22, 2010

JOHN A. QUELCH

HEATHER BECKHAM

# Metabical: Positioning and Communications Strategy for a New Weight-Loss Drug

*"I have tried countless diets and every new weight-loss pill that has come on the market. Nothing seems to take off those extra pounds. With diets, I am miserable because I am starving all the time, and none of the weight-loss pills seem to work. I might lose a couple pounds, but I never reach my weight-loss goals and I usually end up gaining more back. I would give anything to lose this extra 20 pounds, so that I can live a longer, happier life."*

—Tamara Jinkens: focus group participant, age 42

Barbara Printup, senior director of marketing for Cambridge Sciences Pharmaceuticals (CSP), listened as overweight focus group participants recounted their lifelong struggles with weight loss. Printup had just been placed in charge of the upcoming U.S. product launch of CSP's newest prescription drug, Metabical (pronounced Meh-tuh-*bye*-cal). In clinical trials, Metabical proved to be safe and effective in stimulating weight loss for moderately overweight individuals.

CSP was an international health care company with a focus on developing, manufacturing, and marketing products that treat metabolic disorders, gastrointestinal diseases, immune deficiencies, as well as other chronic and acute medical conditions. The company captured over $25 billion in sales in 2007. Printup had over 20 years of experience marketing prescription drugs for CSP. She had led six new drug campaigns and had just concluded work on Zimistat, CSP's most successful product launch to date.

Final FDA approval for Metabical was expected in the coming year, and the product launch was scheduled for January 2009. It was now February 2008, and Printup's first order of business was to develop a viable positioning strategy and associated marketing communications plan for Metabical.

## Overweight Adults in the U.S.

Researchers and health care professionals measure excess weight using the Body Mass Index (BMI) scale. The BMI scale[1], which calculates the relationship between weight and height associated

---

[1] BMI = body weight in kilograms divided by height in meters squared.

HBS Professor John A. Quelch and writer Heather Beckham prepared this case solely as a basis for class discussion and not as an endorsement, a source of primary data, or an illustration of effective or ineffective management. This case, though based on real events, is fictionalized, and any resemblance to actual persons or entities is coincidental. The narration includes occasional references to actual companies.

Copyright © 2010 Harvard Business School Publishing. To order copies or request permission to reproduce materials, call 1-800-545-7685, write Harvard Business Publishing, Boston, MA 02163, or go to http://www.hbsp.harvard.edu. No part of this publication may be reproduced, stored in a retrieval system, used in a spreadsheet, or transmitted in any form or by any means—electronic, mechanical, photocopying, recording, or otherwise—without the permission of Harvard Business Publishing.

Harvard Business Publishing is an affiliate of Harvard Business School.

with body fat and health risk, is appropriate for both men and women. It has three BMI categories of excess weight for adults: overweight (25 to 30); obese (30 to 40); and severely, or morbidly, obese (over 40).

## Health and Social Issues

Excess weight is considered a public health crisis in the U.S., with approximately 65% of the entire adult population categorized as overweight, obese, or severely obese. Being overweight is related to a number of serious health complications, and according to the American Obesity Association in 2005, "the second leading cause of preventable death in the U.S."[2]

In addition to health risks, overweight individuals endure a significant social stigma as well as outright discrimination. Laziness and self-indulgence are common stereotypes associated with this group. Many overweight people feel like social outcasts. The professional life of an overweight individual could also be negatively affected, as excess weight has been found to adversely influence hiring decisions, wages, and promotions.[3]

## Weight-Loss Drugs

No prescription-drug options specifically for the overweight segment (BMI of 25 to 30) were available in 2008. Although a plethora of over-the-counter (OTC) weight-loss drugs existed, only the OTC drug Alli had been approved by the FDA for weight-loss use. Alli was a reduced-strength version of the prescription drug Xenical[4]. Alli users took one pill with each meal. Negative side effects of the drug included gastrointestinal conditions such as loose stools, increased defecation, incontinence, and abdominal pain. These side effects worsened when the patient's diet included too much fat. Other, more-serious side effects of Alli and Xenical were also being investigated. Printup had recently learned that FDA regulators were reviewing over 30 reports of liver damage, including six cases of liver failure, in patients who had taken Alli and Xenical between 1999 and 2008.[5]

All other OTC weight-loss solutions (e.g., hoodia, chromium, green tea extract, conjugated linoleic acid, chitosan, bitter orange, etc.) were categorized as herbal or dietary supplements by the FDA and were, therefore, unregulated by the agency.

The drug industry faced several safety concerns with regard to weight-loss drugs and had been accused of deceptive marketing claims that dampened enthusiasm for the products. Given that herbal remedies and dietary supplements did not require stringent FDA testing and approval, health complications from their use might not be discovered until after the product hit the market. In one high-profile example, the dietary supplement ephedra was linked to several cases of sudden cardiac death and other serious health risks. Consequently, the FDA instituted an outright ban on the purchase or sale of ephedra. Deceptive marketing claims also hurt industry credibility. In early 2007,

---

[2] "American Obesity Association Fact Sheet" (2005, May 2). Retrieved 4/1/10 from American Obesity Society website: http://obesity1.tempdomainname.com/subs/fastfacts/obesity_US.shtml.

[3] Puhl, Rebecca. "Understanding the Stigma of Obesity and Its Consequences." Retrieved 4/1/10, from Obesity Action Coalition website: http://www.obesityaction.org/magazine/oacnews3/Stigma%20of%20Obesity.pdf.

[4] Xenical and all other prescription weight-loss drugs were recommended only for patients with BMIs of 30 or greater.

[5] On August 24, 2009, the FDA announced it was reviewing adverse event reports of liver injury in patients taking the weight-loss drug orlistat, marketed as the prescription drug Xenical and OTC medication Alli. At press time of the case, the FDA's analysis of this data was still ongoing, and no definite association between liver injury and orlistat had been established. "Early Communication about an Ongoing Safety Review Orlistat" (2009, August 24). Retrieved 3/18/10, from FDA website: http://www.fda.gov/Drugs/DrugSafety/PostmarketDrugSafetyInformationforPatientsandProviders/DrugSafetyInformation forHeathcareProfessionals/ucm179166.htm.

the Federal Trade Commission required manufacturers of the popular OTC weight-loss drugs TrimSpa, Xenadrine EFX, CortiSlim, and One-A-Day WeightSmart to pay $25 million to settle allegations that the products' weight-loss claims were unsubstantiated.

CSP believed that its prescription drug, Metabical, was far superior to any weight-loss solution on the market at that time.

## Metabical

CSP's Metabical would be the first prescription drug approved by the FDA specifically for overweight individuals (i.e., those with a BMI of 25 to 30). Researchers at CSP created a drug that combined a new appetite-suppressant compound, calosera, with a revolutionary fat-blocking and calorie-absorption agent, meditonan. The combination of calosera and meditonan produced dramatic weight loss for overweight individuals. It worked in a low-dose formulation, thereby reducing stress on heart or liver functions that other weight-loss drugs tended to produce. Metabical also contained a controlled-release feature that required only one pill be taken per day. Negative side effects of the drug occurred when users consumed high levels of fat and calories. These side effects were similar to the gastrointestinal discomfort caused by Alli, only less severe.

Clinical trials proved Metabical to be effective in achieving significant weight loss for overweight individuals. The majority of trial participants reached their weight-loss goals by week 12. These studies revealed that overweight individuals with BMIs of 28 to 30 lost an average of 26 pounds when taking Metabical, compared with an average loss of 6 pounds for patients within the same BMI range who took a placebo (control group). For participants with BMIs of 25 to 28 weight loss averaged 15 pounds for Metabical users versus an average of 2 pounds for those in the control group.

Because Metabical had a few negative side effects associated with excess fat and calories in the diet, it also helped with behavior modification and healthier eating habits. On average, individuals who took Metabical maintained, for three years, weight-loss levels within 10% of what the clinical trial participants had experienced. Metabical's formulation was not very effective in helping individuals with BMIs of 30 or greater lose weight and was, therefore, not recommended for this group.

Although pricing had not been finalized, CSP estimated the retail price for the drug would be approximately $3 to $5 per day, with the average course of treatment lasting 12 weeks. Many health insurance plans (both HMOs and PPOs) excluded anti-obesity drugs from coverage. The majority of weight-loss drugs were purchased by patients "out of pocket," without reimbursement from health insurance carriers. Initial reports confirmed that few prescription drug plans would cover the cost of Metabical. Printup had been thinking about a campaign to persuade managed health care plans to include Metabical in their prescription-drug programs. However, she planned to review the first six months of sales data before taking that step.

## Support Program

CSP planned to create a comprehensive support program to complement the Metabical pill. Its goal was to enable individuals to achieve better results than they would from the pill alone. In addition, the support program would teach lifestyle skills for healthy weight maintenance after the initial weight loss was achieved.

CSP expected to spend $200,000 on the development of the support program, and Year 1 costs associated with producing the program were expected to be $2 million. In total, the program was estimated to account for just under 10% of the total Year 1 marketing budget. The support program would include reference materials; online weight-control tools (e.g., weight-loss tracker, food diary, nutritional and calorie calculator); personal support (e.g., community forums); meal plans (e.g., menu planner, grocery lists, and thousands of recipes); and exercise plans (e.g., weight-training and cardio routines of less than 20 minutes per day). Although OTC drug Alli also had an online support plan, CSP intended to make its Metabical support program more comprehensive by giving users access to it for 24 months.

The support program had not been included in the extensive FDA clinical trial testing. As a result, studies had not yet measured the impact it would have on the product's efficacy. However, Printup surmised it would significantly enhance the ability of Metabical users to reach and maintain their weight-loss goals. The lingering question was how she would highlight the support program within the communications strategy.

## Market Research

Between 1976 and 2000, the U.S. experienced an alarming rise in the number of overweight and obese adults. By 2000, 34% of the population was considered overweight, another 25.8% were classified as obese, and an additional 4.7% fell into the severely obese category. The percentage of the U.S. adult population that was overweight increased steadily with age for both men and women, with the highest incidence among men age 65 to 74 and women age 55 to 64. The population with less than a high school education had the highest prevalence of obesity. However, excess weight was a problem for all demographic segments. **Exhibit 1** provides a summary of key demographic information.

Health care providers were enthusiastic about the prospect of a drug that could aid in weight loss and help establish healthier diet and exercise habits. Interviews with health care providers revealed one of their top concerns with weight-loss drugs was the likelihood that patients would regain weight once they stopped taking the pills.

One health care provider commented,

> "Many of my moderately overweight patients need help shedding unhealthy pounds. I see cases of diabetes and heart disease each day that are a direct result of being overweight. I have tried, with little success, to counsel these patients with diet and exercise plans. I am pleased that Metabical will provide these patients with assistance to reach a healthy weight. However, taking a pill each day is not a long-term solution. I am impressed with the tools and customized action plans provided in the proposed support program. All this helps with behavior modification and increases the chances that the weight loss will be maintained once the drug regimen is completed."

Results from an extensive marketing survey of overweight individuals commissioned by CSP in 2007 revealed considerable interest from the overweight market as well. Over 70% of the respondents indicated that they were dissatisfied with their current weight. In addition, 35% of respondents were actively trying to lose weight and 15% of the people in that subsegment were comfortable using drugs to help reach their weight-loss goals. When survey participants were asked specifically about a prescription weight-loss drug for overweight individuals, 12% said they would immediately make an appointment with their health care provider and request a prescription. Additional survey results are shown in **Exhibit 2**.

Printup also commissioned a study to analyze psychographic segmentation of overweight individuals. The study revealed a wide variety of attitudes toward physical activity, portion control, food preferences, nutrition, self-image, and overall health. Gender played a significant role in shaping life views on weight and body image. Women demonstrated the most distinct segmentation and clustered around five discrete psychographic profiles. **Exhibit 3** summarizes them.

Furthermore, the February 2008 focus group produced valuable insight into the struggles of overweight individuals. A common theme was the dissatisfaction with current weight-loss options and the desire for a proven and safe way to drop excess weight. Focus group participants expressed the desire for a prescription-strength drug with FDA approval and clinical results to back up weight-loss claims.

## Marketing Communications Strategy

The communications strategy of a prescription weight-loss drug such as Metabical had to address all participants in the decision-making process. Printup's plan focused heavily on both the end consumer (the patient) and the health care providers who would be prescribing the medication. Compliance (e.g., taking the pills each day, refilling prescriptions, and supplementing the drug with a reasonable diet and exercise regimen) was also an important part of Metabical's success, and Printup wanted to make sure her marketing communications strategy addressed these post-purchase activities. The initial Metabical marketing launch budget, shown in **Exhibit 4**, was based on CSP's most recent drug launch. Printup used this as a starting point and intended to fine-tune the estimates to reflect appropriate levels of spending for Metabical. In addition, Printup planned to conduct comprehensive testing of the advertising and promotion campaigns throughout the first year. Using the feedback from this analysis, she would adjust the budget accordingly.

### Advertising

Direct-to-consumer (DTC) advertising was somewhat of a new phenomenon in the drug industry. In 1997, the FDA introduced guidelines that opened the flood gates to the ubiquitous drug advertisements the public would become accustomed to by 2008. Printup's strategy included a DTC television, online, radio, and print media blitz at the time of the drug's launch and heavy advertising throughout the first year to establish the Metabical name. Printup believed that over half of the total Year 1 marketing budget should be dedicated to DTC advertising. Patients' knowledge and awareness of Metabical was her number one priority. CSP's advertising agency had provided three initial concepts for the DTC ads:

- *Losing weight is tough.* You don't have to do it alone. Let Metabical and your health care provider start you on the road to a healthy weight and better life.

- *Look your best.* Shed excess pounds with Metabical and discover a happier, more attractive you. Metabical – all you need to succeed.

- *Those extra 20 pounds could be killing you.* Being overweight leads to heart disease, high blood pressure, diabetes, and gallbladder disease. It's time to get healthy – Metabical can help.

Printup also considered featuring a well-known celebrity endorser, publicly known to have struggled with her weight, in the DTC ads. Celebrity spokespeople were quite common in the weight-control industry (e.g., Valerie Bertinelli for Jenny Craig) and were becoming more mainstream with prescription drugs (e.g., Sally Field for Boniva). In fact, many industry observers credit Sally Field's endorsement of Boniva for the drug's success over the much less expensive generic medication, alendronate, which was proven to be equally effective. Senior executives initially balked at the celebrity spokesperson idea, so the current marketing budget did not include funds for this marketing tactic.

Metabical's advertising strategy also needed to target the professional medical community. This part of the campaign would include print ads in leading medical publications (e.g., *Journal of the American Medical Association*) and interactive ads adjacent to online physician-resource information (e.g., PDR.net) to raise awareness about the drug and its benefits. Printup planned to run the advertising heavily during the three months leading up to the drug's launch and then steadily throughout the first year. This push advertising was much more modest, with Printup estimating it at less than 5% of the total marketing budget. CSP's advertising agency had also developed three different advertising concepts aimed at health care providers:

- Give your overweight patients a safe alternative to fad diets and dangerous OTC drugs. Introducing Metabical – a clinically proven weight-loss drug.

- Atherosclerosis, coronary artery disease, high blood pressure, diabetes, and gallbladder disease. Your overweight patients are dying for help. Introducing Metabical – FDA approved weight-loss drug.

- Empower your patients to lose excess weight, change their unhealthy eating habits, and achieve long-term success. Introducing Metabical – short-term drug therapy and comprehensive support program for overweight patients. It gets results.

*Promotion and Public Relations*

The promotion budget also included campaigns aimed at both the health care providers and the end consumers. Approximately $1.3 million[6] was currently earmarked for these promotional activities in Year 1. CSP's previous promotion strategies had always included a direct mail campaign. Before the Metabical launch, Printup had planned a mailing for 100,000 health care providers, which included an informational pamphlet about the drug and a reply card offering a sample of the support program. In addition, Printup developed a viral marketing campaign, "The Metabical Challenge," which was to coincide with the drug's launch. The Metabical Challenge would be an online contest in which Metabical users would have the chance to compete to see who could reduce their BMIs by the highest percentage. Printup also planned to use social networking sites and contestant blogs to create buzz about Metabical. If done correctly, social and viral marketing had the potential to be an extremely valuable, cost-effective medium. However, Printup had little experience with social media, and she wondered about the most effective way to utilize this mix element and how large a role it should play in the communications strategy.

Public relations efforts would include not only well-timed pre-launch and at-launch press releases, but also two high-profile medical education events and a series of podcasts aimed at physicians. The first event, which was to occur one month before the drug's launch, would be a roundtable discussion involving prominent thought leaders in the medical community. Printup had discussed coverage of this event with several leading news organizations. The second event would be a medical research symposium that was open to members of the media and to medical professionals. This event would occur the same week as the drug's launch. Leading medical researchers in cardiology, diabetes, gastroenterology, depression therapy, and oncology had already committed to serve as keynote speakers, and a series of breakout sessions focusing on health issues for overweight patients were scheduled. Printup estimated approximately $4.3 million would be spent in total on public relations activities in the first year.

---

[6] This figure does not include support program development or production costs.

*Sales Force*

A CSP sales team responsible for two gastrointestinal drugs was directed to detail Metabical to health care providers in their territories and add it to their existing portfolios. Printup planned to work with the director of sales for this team to develop sales scripts and presentations providing clinical information. For established drugs, CSP sales reps typically visited targeted medical offices four times a year to discuss the drug and provide samples. In the past, CSP sales reps who sponsored promotional Lunch & Learn seminars for health care providers received great feedback. These lunchtime presentations focused on information about the featured drug. CSP sales reps believed that health care providers and office staff usually came for the free food, but it was a great opportunity to learn more about the providers' practices and to share information about the drug in a relaxed setting. Printup still had to decide how she would best utilize the sales force that had been allocated to Metabical.

The Metabical sales team consisted of 32 sales representatives who called on approximately 3,200 medical offices. These reps focused on offices that were the most geographically accessible and responsive to drug-rep visits. Each CSP sales representative visited four practices per day, on average.

## Conclusion

Printup sat in her office, staring at a blank legal pad. At the top, she circled and underlined the figures "10 years" and "$400 million." These represented how much time and money CSP had spent in R&D and on FDA trials for Metabical. Printup was well aware that in order to recoup this massive investment, the drug needed not only a successful launch, but also long-term, steady demand. Although all the medical studies and consumer research showed great promise for Metabical, Printup knew that poor positioning of the drug could spell disaster. If Metabical were not successful with initial consumers, credibility of the drug would be in question and FDA approval would mean little. CSP would have about 10 years of exclusivity once Metabical hit the market before generic versions could be manufactured. Ensuring the longevity of Metabical and avoiding being pigeonholed as a fad-diet cure were at the top of Printup's agenda. Printup had been vying for a coveted VP position in CSP's corporate marketing group. She was told that if the Metabical launch were successful, she would be rewarded with this promotion.

Printup still needed to flesh out the optimal segmentation, targeting, and positioning of the drug. Then she could then move on to assessing her current marketing communications strategy and developing a timeline for the key activities. As Printup began to pore over the data she had gathered, questions kept popping into her mind. Who was the ideal target consumer? How should each participant in the decision-making process be addressed? How could these participants best be reached? What was the appropriate message to convey to each one of them? What was the role of the support program? What was the optimal rollout schedule for key marketing communications activities?

**Exhibit 1** Trends: Percentage of Overweight, Obese, and Severely Obese Adults in the United States, 1976–2001[a]

|  | % of U.S. Adults | | | U.S. Adult Population |
|---|---|---|---|---|
|  | Overweight ($25 \leq BMI < 30$) | Obese ($30 \leq BMI < 40$) | Severely Obese ($BMI \geq 40$) | (millions)[b] |
| 1999 to 2000 | 34.0 | 25.8 | 4.7 | 209 in 2000 |
| 1988 to 1994 | 33.0 | 20.1 | 2.9 | 185 in 1990 |
| 1976 to 1980 | 31.6 | 14.4 | No Data | 163 in 1980 |

|  | 1999–2000 Men ($BMI \geq 25$) Prevalence (%) | Women ($BMI \geq 25$) Prevalence (%) |
|---|---|---|
| Overall | 67.0 | 62.0 |
| Age (years) | | |
| 20 to 34 | 58.0 | 51.5 |
| 35 to 44 | 67.6 | 63.6 |
| 45 to 54 | 71.3 | 64.7 |
| 55 to 64 | 72.5 | 73.1 |
| 65 to 74 | 77.2 | 70.1 |
| 75+ | 66.4 | 59.6 |

| Education Level | 2001 Obesity (%) | Income Level | 2001 Obesity (%) |
|---|---|---|---|
| Less than High School | 27.4 | Less than $25,000 | 32.5 |
| High School | 23.2 | $25,000–$40,000 | 31.3 |
| Some College | 21.0 | $40,000–$60,000 | 30.3 |
| College | 15.7 | More than $60,000 | 26.8 |

[a] "American Obesity Association Fact Sheet" (2005, May 2). Retrieved 3/18/10 from American Obesity Society website: http://obesity1.tempdomainname.com/subs/fastfacts/obesity_US.shtml

Hitti, Miranda (2005, May 2). "Rich-Poor Gap Narrowing in Obesity." Retrieved 3/18/10, from WebMD website: http://www.webmd.com/diet/news/20050502/rich-poor-gap-narrowing-in-obesity

*Statistical Abstract of the United States: 2006.* Retrieved 3/18/10 from U.S. Census Bureau website: http://www.census.gov/prod/2005pubs/06statab/pop.pdf

[b] The size of the U.S. adult population was approximately 230 million in 2008.

**Exhibit 2**  Selected Results: 2007 Marketing Survey[a]

- 70% of the respondents surveyed were not satisfied with their current weight.

- 35% of the respondents surveyed were actively trying to lose weight.

- 15% of those actively trying to lose weight were comfortable using drugs to help reach their weight-loss goals.

- When the features and benefits of Metabical were described to respondents, 12% of respondents surveyed said they would immediately make an appointment with their health care provider and request a prescription.

- 75% of women and 65% of men surveyed were dissatisfied with their current weight and appearance.

- 50% of women and 30% of men surveyed visited a health care provider for a yearly physical exam.

- 55% of women and 40% of men surveyed said they wanted to change their behavior to live a healthy lifestyle.

- 60% of women and 30% of men surveyed had tried and failed to lose weight in the past five years.

- 65% of women and 35% of men surveyed were dissatisfied with current weight-loss options on the market.

- 75% of respondents surveyed with a college degree and 45% of respondents surveyed with a high school diploma were aware of the health risks associated with being moderately overweight.

- 65% of respondents surveyed age 18 to 35 and 40% of respondents surveyed age 35+ said they wanted to lose weight to look better; 35% of respondents surveyed age 18 to 35 and 60% of respondents surveyed age 35+ said they wanted to lose weight improve their overall health.

- 5% of respondents surveyed with incomes less than $40,000, 11% of respondents surveyed with incomes between $40,000 and $80,000, and 20% of respondents surveyed with incomes over $80,000 stated that they would be willing to pay "out of pocket" for a prescription weight-loss drug.

[a] Survey included 1,000 men and 1,000 women, age 18 to 70 with BMIs of 25 to 29.9, from various socioeconomic levels.

**Exhibit 3** Female Psychographic Segmentation Report Summary

| Segment | Description | Typical Demographic Profile |
| --- | --- | --- |
| "I want to look like a movie star" | Fixated on body image and achieving the perfect physique. Low self-esteem and unrealistic expectations. | Age 18 to 30, high school education, household income under $40,000 |
| "I want to be healthier" | Want to lose weight to feel better and live longer. Knowledgeable about the importance of nutrition and exercise. Ready to make a change. | Age 35 to 65, college education plus, household income $80,000+ |
| "I want to wear my skinny jeans" | Focused on goal of reclaiming former weight. Motivated and willing to alter current behavior. | Age 25 to 40, college education, household income $50,000–$80,000 |
| "I want to lose weight, but only if it is easy" | Don't want to be deprived of indulgences. Not interested in changing diet or exercise habits. | Age 45 to 65, some college education, household income $40,000–$60,000 |
| "I am fine the way I am" | Don't see need for change. In denial about negative health consequences associated with being overweight. | Age 40 to 65, some college education, household income $30,000–$50,000 |

Exhibit 4    Metabical First-Year U.S. Marketing Budget (thousands of dollars)

|  | Year 1 |
|---|---|
| **Advertising** | |
| Push (prescriber) | $ 1,000,000 |
| Pull (direct-to-consumer) | $ 12,000,000 |
| **Total Advertising** | $ 13,000,000 |
| **Promotion** | |
| Development of support program | $ 200,000 |
| Lunch & Learn seminars/Other promo | $ 600,000 |
| Production of support program | $ 2,000,000 |
| Training/promotional materials | $ 500,000 |
| Direct mailings to health care providers | $ 200,000 |
| **Total Promotion** | $ 3,500,000 |
| **Public Relations** | |
| Medical education meetings and events | $ 3,500,000 |
| Press release/materials | $ 800,000 |
| **Total Public Relations** | $ 4,300,000 |
| Market research | $ 600,000 |
| Sales force allocation | $ 1,491,000 |
| Product management allocation | $ 255,000 |
| **Total Budget** | $ 23,146,000 |

HARVARD | BUSINESS | SCHOOL

**BRIEF CASES**

2069
MAY 28, 2007

HEIDE ABELLI

# Mountain Man Brewing Company: Bringing the Brand to Light

It was February 20, 2006, in the New River coal region of West Virginia. Chris Prangel, a recent MBA graduate, had returned home a year earlier to manage the marketing operations of the Mountain Man Beer Company (MMBC), a family-owned business he stood to inherit in five years, when his father, Oscar Prangel, the president and owner, retired. Mountain Man brewed one beer, *Mountain Man Lager*, also known as "West Virginia's beer."

Due to changes in beer drinkers' preferences, the company was now experiencing declining sales for the first time in the company's history. In response, Chris wanted to launch *Mountain Man Light*, a "light beer" formulation of *Mountain Man Lager*, in the hope of attracting younger drinkers to the brand. Over the previous six years, light beer sales in the United States had been growing at a compound annual rate of 4%, while traditional premium beer sales had declined annually by the same percentage. Earlier that day, Chris met with a regional advertising agency about a marketing campaign to launch *Mountain Man Light*. Back in his office, he watched an agency videotape from a focus group. He observed a half-dozen participants, 21 to 55 years old, showing various reactions to proposals to extend the Mountain Man brand to a new light beer product.

- A man in his fifties leaned into the facilitator and declared, "*Mountain Man Light*? Come on, I'm not interested in light beer. Just don't mess with *Mountain Man Lager*."

- A man in his early thirties, dressed in jeans and a camouflage shirt, stared at a mock advertisement and shouted, "Fancy barbecue parties, with puppies running around.... What do they have to do with Mountain Man?"

- A man, in his mid-twenties and fashionably-dressed, said, "Sounds pretty corporate... I think the beer is too strong for me anyway. I'll leave it to these guys to drink."

- A woman in her early twenties wearing low-rise jeans and a trendy T-shirt commented, "Mountain Man is kind of 'retro cool.' I like light beer and Miller Lite is so passé. I would definitely try *Mountain Man Light*."

---

Heide Abelli prepared this case solely as a basis for class discussion, and not as an endorsement, a source of primary data, or an illustration of effective or ineffective management. Heide Abelli is a former consultant with the Monitor Group, a strategy consulting firm based in Cambridge, MA, where she consulted to a variety of consumer products companies on marketing issues. The author thanks the following executives from the brewing industry; their help was indispensable in refining the case: Brent Ryan of Newport Storm, Rob Schimony of Yuengling & Son, and Charlie Storey of Harpoon Brewery.

This case, though based on real events, is fictionalized, and any resemblance to actual persons or entities is coincidental. There are occasional references to actual companies in the narration.

Copyright © 2007 President and Fellows of Harvard College. To order copies or request permission to reproduce materials, call 1-800-545-7685, write Harvard Business Publishing, Boston, MA 02163, or go to http://www.hbsp.harvard.edu. This publication may not be digitized, photocopied, or otherwise reproduced, posted, or transmitted, without the permission of Harvard Business School.

Chris switched off the videotape and glanced up at a photograph of his father with a group of rugged, middle-aged men from the Coal Miner's Union. Although Chris firmly believed that the window of opportunity for introducing *Mountain Man Light* was closing, Oscar had warned, "Look at what new product lines get you… 90% more products, 90% more chance you'll kill your core brand." Chris wondered how the men in the photograph would react to a billboard picture of yuppies consuming *Mountain Man Light*. Could Mountain Man command as much pride for the brand from his generation as it had from his father's? Moreover, could he reposition the brand to drive sales of *Mountain Man Light* to young people without eroding the core brand equity of *Mountain Man Lager*? As Chris prepared to discuss the brand extension with Oscar, he knew that whatever strategy Mountain Man pursued, it would have dramatic implications for the brand, the company, and his family.

## Mountain Man: The Company and the Brand

Guntar Prangel founded the Mountain Man Beer Company (MMBC) in 1925. Mr. Prangel had reformulated an old family brew recipe using a meticulous selection of rare, Bavarian hops and unusual strains of barley, resulting in a flavorful, bitter-tasting beer which the Prangel family launched as *Mountain Man Lager*. By the 1960s, *Mountain Man Lager's* reputation as a quality beer was well entrenched throughout the East Central region of the United States.[1]

*Mountain Man Lager* was a legacy brew in a mature business. By 2005 Mountain Man was generating revenues just over $50 million and selling over 520,000 barrels[2] of *Mountain Man Lager* beer primarily to distributors in Illinois, Indiana, Michigan, Ohio, and its native West Virginia. (See **Exhibit 1** for MMBC income statement.) It had held the top market position among lagers in West Virginia for almost 50 years and had respectable market share for an old school, regional brewery in most of the states where the beer was distributed. To accentuate the beer's dark color, it was packaged in a brown bottle, with its original 1925 design of a crew of coal miners printed on the front. *Mountain Man Lager* was priced similarly to premium domestic brands such as Miller and Budweiser and below specialty brands such as Sam Adams. Its price was typically $2.25 for a 12-ounce serving of draft beer in a bar and $4.99 for a six-pack in a local convenience store.

Brand played a critical role in the beer-purchasing decision. When selecting beer, consumers considered several factors: taste; price; the occasion being celebrated; perceived quality; brand image; tradition; and local authenticity. MMBC relied on its history and its status as an independent, family-owned brewery to create an aura of authenticity and to position the beer with its core drinkers—blue-collar, middle-to-lower income men over age 45. (**Exhibit 2** provides profiles of the average *Mountain Man Lager* consumer in contrast to average profiles of premium-beer and light-beer drinkers.) In a recent study in West Virginia, this audience had rated *Mountain Man Lager* as the best-known regional beer, with an unaided response rate of 67% from the state's adult population. In 2005, Mountain Man Lager won "Best Beer in West Virginia" for its eighth year straight (it also won "Best Beer in Indiana") and was selected as "America's Championship Lager" at the American Beer Championship.

Brand awareness was one cornerstone of the brand's success with blue-collar consumers. Market research showed that Mountain Man was as recognizable a brand among working-class males in the East Central region as Chevrolet and John Deere. The other cornerstones were the perception of quality in *Mountain Man Lager* and the brand loyalty it cultivated. There were ranges of subjective

---

[1] The East Central beer region of the United States consisted of seven states: Illinois, Indiana, Kentucky, Michigan, Ohio, West Virginia, and Wisconsin.

[2] One beer barrel = 31 U.S. gallons = 2 "half-barrel" (15.5 gallon) kegs = 13.78 cases (of 24 12-ounce bottles).

attributes that defined the quality of Mountain Man, like its smoothness, percentage of water content, and "drinkability"—but it was *Mountain Man Lager's* distinctively bitter flavor and slightly higher-than-average alcohol content that uniquely contributed to the company's brand equity. One participant in the recent focus group seemed to have spoken for many customers: "My dad drank Mountain Man just like my granddad did. They both felt it was as good a beer as you could get anywhere."

Over the years, MMBC had invested in a number of branding activities to build "brand equity" with core consumers. Mountain Man's distributors also handled Anheuser Busch and numerous specialty beer products. Because these distributors tended to focus on servicing their main customer, they would not reliably strive to build Mountain Man's brand. MMBC therefore established its own small sales force, which didn't just help push the brand; it proselytized, focusing on one ultimate objective: getting off-premise locations (like liquor stores or supermarkets) to embrace Mountain Man. Blue-collar males purchased 60% of the beer they drank at off-premise locations. Mountain Man sold 70% of its beer for off-premise (liquor stores) consumption, consistent with average industry sales through this channel.

## Mountain Man's Competition

The competition in the U.S. beer market fell into four categories: Major and second-tier domestic producers, import beer companies, and specialty brewers.

*Major domestic producers* consisted of a handful of companies who competed on the basis of economies of scale in production and advertising. This highly concentrated segment of the market was dominated by three companies: Anheuser Busch, Miller Brewing Company, and Adolf Coors Company. Together, these companies accounted for 74% of 2005 beer shipments in Mountain Man's region.

*Second-tier domestic producers* consisted of medium-sized competitors, such as Pabst Brewing Company and Genessee which, similar to the major domestic producers, sold their beers nationally to distributors and retailers. In addition, there were smaller, regional players that produced between 15,000 and two million barrels of beer per year and generally limited distribution to areas surrounding their plants, selling their beer to regional distributors and retailers. By November 2005, there were roughly 30 regional breweries in the United States. These companies followed the same product and marketing strategy as the major domestic producers, but lacked the financial and marketing resources to defend their brands as aggressively. The second-tier domestic producers accounted for 12.5% of beer shipments in the East Central region in 2005.

*Import beer companies* from Germany (Beck's, for example), Holland (Heineken), Canada (Molson), and Mexico (Corona) traditionally served the needs of sophisticated beer drinkers who desired more flavorful, bitter-tasting beer products. They operated at a distinct disadvantage relative to domestic competitors due to higher shipping costs, weaker distribution networks, an inability to control product freshness, and margin reduction due to weakening of the U.S. dollar. In 2005, import companies controlled about 12% of the region's market.

*The craft beer industry* was divided into four markets: brewpubs, microbreweries, contract breweries, and regional craft breweries. They all brewed beer using traditional malt ingredients, were independently owned, and by definition produced less than two million barrels annually. *Brewpubs* were restaurant/bar establishments with over 25% of their beer products brewed and consumed on site. In 2005, more than 980 brewpubs operated in the United States, accounting for 10% of the craft brew volume. *Microbreweries* traditionally operated in limited distribution networks and produced less than 15,000 barrels a year. In 2005, the 380 U.S. microbreweries accounted for 12% of the craft beer volume. *Contract breweries*, breweries that manufactured beer for client firms, accounted for 16% of the craft beer volume. Finally, almost 50 U.S. *regional craft brewers* (such as Sam Adams, Sierra Nevada, and Harpoon), producing more than 15,000 barrels annually, accounted for the remaining 62% of the market. In the East Central region, all craft brewers together controlled 1.5% of the total beer market. (See **Exhibit 3** for competitive market shares by brewer type in the East Central region.)

## The Situation at Mountain Man in 2005

The United States was the largest beer-consuming market in the world, with over $75 billion in annual sales in 2005. Since 2001, U.S. per capita beer consumption had declined by 2.3%, largely due to competition from wine and spirits-based drinks, an increase in the federal excise tax, initiatives encouraging moderation and personal responsibility, and increasing health concerns.

Of total U.S. beer sales, 18.3% took place in the East Central region. (See **Exhibit 4** for East Central beer consumption overall and by state.) Although imports and craft beers didn't have quite the stronghold in the "heartland" states (where MMBC sold its beer) as they did in other parts of the country, even there, both categories were beginning to take hold. Some states in the region, including West Virginia, had become particularly competitive; the state had recently repealed arcane laws that had sharply limited the promotion of beer in retail establishments, and as a result, retail stores began selling beer at deep discounts. Distributors became more discriminating about which smaller brands they would continue to carry, paying more attention to turnover and margins, and dropping brands that contributed little to the bottom line. Large national brewers, who maintained economies of scale in brewing, transportation, and marketing, put great pressure on the smaller, regional breweries like Mountain Man.

This pressure, combined with a glut of product, led to the closing of many independent breweries in the East Central region over the past 40 years. Breweries that once reigned supreme across the region had disappeared, taking with them the loyal allegiance of their communities. MMBC's survival was in large part due to the fact that it served a large enough market with a very strong brand, and it therefore could continue to compete against national players with deep pockets such as Anheuser Busch, the company's most significant competitor. There were only four breweries left in West Virginia by 2005, and Mountain Man's 2005 revenues were down 2% relative to the prior fiscal year. Even though the company was still profitable in spite of the sales decline, the prospect of continued downward pressure on revenue would challenge the company's ability to remain profitable. Facing an aging demographic in the shrinking premium segment of the beer market, the company struggled to maintain a steady share of its market segment against the large domestic brewers, which were spending heavily to maintain their own sales levels in the premium segment.

Beer was not subject to sharp fluctuations in demand during economic downturns. Changes in volume were driven primarily by changes in consumer segments. Most industry observers agreed that the key consumer segment for beer companies was younger drinkers, 21–27 years of age. This group represented the "first-time drinker demographic" that had not yet established loyalty to any particular brand of beer. The segment represented about 13% of the adult population in 2005, but accounted for more than 27% of total beer consumption and was growing. In addition, this age group spent twice as much per capita on alcoholic beverages than consumers over 35 years of age and was forecasted to grow by nearly four million by the year 2010.

Another significant trend was growth in the "light" beer category which had been steadily gaining in market share and accounted for 50.4% of volume sales in 2005, compared with 29.8% in

2001. (See **Exhibit 5** for a breakdown of the East Central regional market by type of beer, and **Exhibit 6** for light beer market shares in the region.) In fact, younger consumers preferred light beer to other categories. They also typically consumed in quantity. However, they tended to buy mainstream brands. A consumer study revealed that while Mountain Man rated high in terms of awareness with the younger, light-beer drinking segment of the market, *Mountain Man Lager* tracked very low as a purchasing preference—as did other lagers and fuller-flavor brews.

Industry observers believed new products introduced beer drinkers to both styles of beer while simultaneously keeping them in the "brand" family. Product line extensions leveraging the core brand name often helped brewers obtain greater shelf space for products and created greater product focus among distributors and retailers. Mountain Man was now alone among the major and regional beer companies in not having expanded its product line beyond its flagship lager product.

In light of these developments, Mountain Man engaged a market research firm to evaluate its single-brand product strategy and brand extension opportunities. The study yielded three interesting findings:

1. *Mountain Man Lager* was known as "West Virginia's Beer." Authenticity, quality, and a unique West Virginia "toughness" were core attributes of the brand. Younger beer drinkers were well aware of the brand, yet perceived the beer as "strong" and a "working man's" beer largely consumed by the "swing" and baby boomer generations. Because younger beer drinkers held "anti-big-business" values, they did show some appreciation for the brand's association with an independent brewery.

2. Traditional advertising was not as effective as grass-roots marketing[3] in building beer brand awareness in certain states in the East Central region, such as West Virginia and Kentucky. Mountain Man had always relied on grass-roots marketing to spread its beer quality message by word of mouth. In contrast, national beer brands used lifestyle advertisements to reach young drinkers. Broadcast spending for beer ads topped $700 million annually, representing over 70% of total advertising expenditures on alcohol. (See **Exhibit 7** for U.S. advertising spending on beer.)

3. A small percentage of MMBC's blue-collar customers accounted for a large percentage of sales, and those customers tended to be very loyal to *Mountain Man Lager*. In fact, the sole brand loyalty rate[4] for *Mountain Man Lager* was 53%, which was higher than the rates of competitive product (i.e., 42% for Budweiser and 36% for Bud Light.) The non-loyal *Mountain Man Lager* customers occasionally spread their consumption across up to five other beer brands.

## The Challenges Ahead at Mountain Man

Chris Prangel pondered the findings of the study. To him it was clear that product preferences in the beer market were changing, and that a light beer product was strategically important to MMBC's future. First, light beer was a newer, fast-growing product category and the only beer category demonstrating consistent growth. Moreover, a light beer would help MMBC gain share in on-premise locations: restaurants and bars. Light beers appealed to younger drinkers overall, and to women, both groups that frequented these locations. Market research indicated that Mountain Man's

---

[3] Grass-roots marketing campaigns typically involve local marketing activities that concentrate on getting as close and personally relevant to individual customers as possible.

[4] The sole brand loyalty rate refers to the percentage of brand users who are loyal to a particular brand and not interested in experimenting with other brands / products.

core customers did not state a brand preference in restaurants and bars. Chris believed Mountain Man's brand recognition could translate into a meaningful share of the local light beer market and hoped that in turn, *Mountain Man Light's* popularity could boost the sales of *Mountain Man Lager*.

Others on the MMBC management team did not share Chris's enthusiasm for launching *Mountain Man Light*. Stretching the brand to target younger drinkers who consumed light beer had dramatic branding implications, not to mention competitive ones. Younger drinkers mirrored the target market for the large national and regional brands. In addition, Oscar Prangel was concerned that launching *Mountain Man Light* would alienate the core customer base and ultimately erode and dilute the Mountain Man brand equity. He was also worried that *Mountain Man Light* might cannibalize the sales of *Mountain Man Lager* because of a fear that retailers would not grant Mountain Man incremental shelf space and therefore would substitute cases of light product for the lager product.

In his last lecture, Oscar had said, "Chris, value is achieved by focusing on what you do best, not by attaching your brand to every conceivable version of a product. Another product line will just add to our cost structure—more inventory, more packaging, more SG&A. We won't sell more barrels; we'll just reduce our profit. Then there's the real risk that *Mountain Man Light* might just end up hurting the sales of *Mountain Man Lager*; I reckon we could count on at least 5% but it might be 20% or higher. Look at how many light beers there are, with millions of dollars invested by their brands. Have they increased the total sales of beer?" To address his father's concerns, sales of *Mountain Man Light* would have to compensate for this potential loss of lager product revenue. However, while Chris understood his father's concerns, he believed that there was a chance that the launch of *Mountain Man Light* might just give *Mountain Man Lager* a lift. He had replied to his father, "This is our chance to play in the light beer sandbox but stay true to the Mountain Man brand by playing on the strengths of our core product."

Chris also wondered if MMBC could afford to launch *Mountain Man Light*. Although the launch of *Mountain Man Light* would not require capital expenditures in plant and equipment in the short term due to existing excess capacity in Mountain Man's facility, launching a new product was an expensive endeavor for a lean company not used to making these kinds of investments. While this was not the launch of a new national beer brand, which Chris knew cost between $10 million and $20 million in TV advertising alone, it wasn't cheap to launch a new product on a regional basis either. The advertising agency estimated that creating a brand awareness level of 60% for *Mountain Man Light* in the East Central region would cost at least $750,000 in an intensive six-month advertising campaign. This would be on top of the $900,000 in annual, incremental SG&A costs that Chris projected the new product would require, based on the need for a *Mountain Man Light* product manager, an addition to the sales staff, and ongoing marketing expenditures. Although MMBC's variable cost per barrel of its lager beer was $66.93, it would cost $4.69 more per barrel to produce *Mountain Man Light*. Because Mountain Man would receive the same price per barrel for both products, the contribution margin for *Mountain Man Light* would be lower than the contribution margin of *Mountain Man Lager*. Chris knew that, given Mountain Man's CFO's conservative stance regarding investment, he would have to convince the senior management team that the *Mountain Man Light* product would generate a profit within two years, selling enough barrels to cover both the associated launch marketing and incremental SG&A expenses and make up for the negative impact on overall profitability resulting from potential lost *Mountain Man Lager* sales. His estimates regarding barrel sales would need to make sense in terms of market share in the very competitive light beer segment. Chris recalled the risks expressed by John Fader, the vice president of sales:

"*Mountain Man Light* will never achieve the volume of larger light beer brands like Miller Lite or Coors Light; those brewers sustain distribution and support advertising in ways we can't. What's more, the big companies are constantly throwing new products bearing the established brand name into the marketplace. *Mountain Man Light* would get lost in that sea of new-product introductions. You'd be doing well if you grabbed a quarter point of market share. We won't get our retailers to give us more facings,[5] so *Mountain Man Light* would just replace facings we have earned for *Mountain Man Lager*. The light beer would only draw time, resources, and attention away from our lager—our bread and butter. Boosting sales of our core brand even slightly means more than what we will get in the light beer segment. It's a pipedream, Chris."

## Chris's Decision

Chris looked at some revenue and net profit projections he had developed to 2010 assuming that *Mountain Man Lager* lost 2% of its revenue base annually. He felt a knot in his stomach as he pondered the "status quo" strategy. He then examined the financial projections he had done a few weeks prior for the *Mountain Man Light* launch, which showed regional revenue growth of the light beer product at 4% annually and Mountain Man steadily growing its share of the regional light beer market by a quarter of a percent each year off of a 2006 base market share of 0.25%.[6]

However, before presenting a formal plan to launch *Mountain Man Light* to his father, Chris needed to think further, strategically and tactically, about marketing and distribution to a new customer segment. How would he address his father's concern that targeting this segment would alienate existing Mountain Man customers and erode core brand equity? What about his father's belief that the Mountain Man brand would never capture the same loyalty among younger beer drinkers that it had from blue-collar workers? Since MMBC did not have the resources to match the marketing efforts of the large, national, light beer brewers, Chris wondered how he would argue that Mountain Man *could* compete against deep-pocketed competitors for the segment. Was he overly optimistic in his projections of the percentage of the light beer market that *Mountain Man Light* could capture?

Chris thought back on what his father had recently said to him, "Chris, I try to keep in mind all the other regional breweries that have vanished over the past 30 or 40 years—Neuweiler, Horlacher, dozens and dozens of them—all gone. I've watched the giants in this business taken down by fatal decisions made at the top that irreversibly damaged the brand. In the '50s, Schlitz sold more barrels than any other brewer. You can't buy a Schlitz beer today. Mountain Man is still standing because we manufacture an exceptional beer with a great brand name, we've never lost sight of our core customer, and we've never been seduced by the other guy's market." Chris valued his father's words and he did not want to be the one to lead Mountain Man down the path to oblivion; however, Mountain Man's revenues were down, and Chris needed to help his father secure the company's future. He wondered if he was missing something or maybe if there were other options he needed to consider. The knot in his stomach tightened again. This was West Virginia's beer. It was the Prangel family beer. It was Chris's legacy staring him in the face.

---

[5] Facings are spaces on the retail shelf, typically in coolers with glass doors.

[6] For purposes of financial analysis in this case, assume Mountain Man's discount rate for evaluating investment projects was 12%.

**Exhibit 1** Mountain Man 2005 Income Statement

| | | |
|---|---:|---:|
| **Net Revenues** | $50,440,000 | 100.0% |
| COGS | 34,803,600 | 69.0% |
| **Gross Margin** | 15,636,400 | 31.0% |
| SG&A[a] | 9,583,600 | 19.0% |
| Other Operating Expenses | 1,412,320 | 2.8% |
| **Operating Margin** | 4,640,480 | 9.2% |
| Other Income | 151,320 | 0.3% |
| **Net Income Before Taxes** | 4,791,800 | 9.5% |
| Provision for Income Taxes | 1,677,130 | 3.3% |
| **Net Income After Taxes** | $ 3,114,670 | 6.2% |

[a] Advertising expenses were $1.35 million annually or 2.7% of total revenues.

Advertising expenses included radio, print, and outdoor advertising, sponsorships, as well as costs to produce these media.

**Exhibit 2** Profile of Beer Drinkers by Beer Type by Key Demographics, 2005

| | Domestic Light Beer | Domestic Premium Beer | Mountain Man Lager |
|---|---|---|---|
| **Gender** | | | |
| Male | 58% | 68% | 81% |
| Female | 42% | 32% | 19% |
| **Age** | | | |
| 21-24 | 9% | 8% | 2% |
| 25-34 | 20% | 20% | 15% |
| 35-44 | 24% | 23% | 19% |
| 45-54 | 22% | 23% | 32% |
| 55-64 | 14% | 14% | 19% |
| 65+ | 12% | 12% | 13% |
| **Household Income** | | | |
| under $25k | 14% | 16% | 20% |
| $25k-49.9k | 25% | 24% | 27% |
| $50k-74.9k | 21% | 21% | 25% |
| $75k-99.9k | 16% | 15% | 15% |
| $100k+ | 24% | 23% | 13% |

Source: First two columns of data extracted from Mintel/Simmons NCS 2005 report, figure 67

**Exhibit 3** Competitive Market Shares in Barrels by Brewer

|  | East Central Region | |
|---|---|---|
| Anheuser-Busch | 15,620,252 | 42.0% |
| Miller | 8,553,948 | 23.0% |
| Coors | 3,347,197 | 9.0% |
| Other 2nd tier Premium & Popular Brewers | 4,648,885 | 12.5% |
| Craft/Specialty Brewers | 557,866 | 1.5% |
| Imports | 4,462,929 | 12.0% |
| Total | 37,191,077 | 100% |

Note: Sales in barrels of wholesale shipments.

**Exhibit 4** Beer Consumption by State, 2000 to 2005 (shipments in barrels)

| STATE | 2000 | 2001 | 2002 | 2003 | 2004 | 2005 |
|---|---|---|---|---|---|---|
| Illinois | 9,038,323 | 9,165,381 | 9,268,188 | 9,108,157 | 9,032,851 | 9,063,267 |
| Indiana | 3,954,209 | 3,947,446 | 4,021,685 | 3,905,265 | 3,993,643 | 3,998,855 |
| Kentucky | 2,517,894 | 2,486,731 | 2,564,013 | 2,490,928 | 2,591,949 | 2,555,739 |
| Michigan | 6,761,561 | 6,695,665 | 6,854,064 | 6,774,702 | 6,746,578 | 6,700,174 |
| Ohio | 8,493,144 | 8,601,604 | 8,682,331 | 8,760,272 | 8,702,382 | 8,584,283 |
| West Virginia | 1,274,626 | 1,311,838 | 1,360,589 | 1,348,527 | 1,373,205 | 1,359,231 |
| Wisconsin | 4,741,019 | 4,784,791 | 4,890,122 | 4,855,313 | 4,877,662 | 4,929,529 |
| East Central Region | 36,780,776 | 36,993,456 | 37,640,992 | 37,243,163 | 37,318,269 | 37,191,077 |
| TOTAL U.S. | 197,609,645 | 200,146,800 | 202,605,792 | 202,586,016 | 204,318,220 | 203,515,148 |

Source: Beer Institute data.

**Exhibit 5** Consumption by Type of Beer and by Origin/Packaging, 2005

A. Consumption by Type of Beer

|  | EAST CENTRAL REGION | % Total | 6-year CAGR |
|---|---|---|---|
| Light Beer | 18,744,303 | 50.4% | +4% |
| Premium Beer | 7,326,642 | 19.7% | (4%) |
| Popular | 4,351,356 | 11.7% | (5%) |
| Imported Premium | 4,462,929 | 12.0% | +6% |
| Superpremium (craft and high-end domestics) | 2,305,847 | 6.2% | +9% |
| Total Barrels | 37,191,077 | 100.0% | |

B. Consumption by Origin/Packaging

|  | EAST CENTRAL REGION | % Total |
|---|---|---|
| Imported | 4,462,929 | 12.0% |
| Domestic - Packaged | 29,618,974 | 79.6% |
| Domestic - Draught | 3,109,174 | 8.4% |
| Total Barrels | 37,191,077 | 100.0% |

**Exhibit 6** Light Beer Market Shares and Dominant Brands

A.

| Light Beer Competitive Market Shares | |
|---|---|
| **East Central Region** | **2005 Market Share** |
| Anheuser-Busch | 49% |
| Miller | 24% |
| Coors | 11% |
| Other brands | 14% |
| Imports | 2% |
| Total Light Beer | 100% |

B.

| Leading Domestic Light Beer Brands | |
|---|---|
| **East Central Region** | **2005 Market Share** |
| Bud Light | 32.9% |
| Miller Lite | 17.8% |
| Coors Light | 14.7% |
| Natural Light | 9.8% |
| Busch Light | 6.4% |
| Michelob Ultra | 5.6% |
| Milwaukee's Best Light | 3.4% |
| Other domestic brands | 9.4% |
| Total | 100% |

C.

| Leading Imported Light Beer Brands | |
|---|---|
| **Brand** | **2005 Market Share** |
| Corona Light | 57% |
| Amstel Light | 26% |
| Labatt Blue Light | 15% |
| Other imported brands | 2% |
| Total | 100% |

Note: Market share calculations based on wholesale barrel sales.

**Exhibit 7** U.S. Beer Advertising Expenditures by Medium (in millions of dollars), 2005

| Medium | 2005 |
|---|---|
| Network television | $382.3 |
| Cable television | 72.1 |
| Spot television | 144.3 |
| Syndicated television | 5.5 |
| Spot radio | 22.4 |
| Network radio | 1.2 |
| **Total Broadcast** | **$627.8** |
| Magazines | 23.2 |
| Newspapers | 6.6 |
| Newspaper supplements | – |
| Outdoor | 51.5 |
| **Total Print** | **$81.3** |
| **TOTAL** | **$709.1** |

# HARVARD | BUSINESS | SCHOOL

9-504-028
REV: JUNE 11, 2007

GAIL MCGOVERN

# Virgin Mobile USA: Pricing for the Very First Time

*When Richard Branson called me to discuss the CEO position at Virgin Mobile USA, I quickly considered the opportunity: a chance to be the chief executive of a newly formed start-up in an overcrowded, increasingly mature, capital-intensive, highly competitive industry. Oh yeah, I should also mention that this is not an industry known for its customer service and we'd be entering with a brand that had little U.S. name recognition except for possibly as an airline. But then I thought, "It's these kinds of opportunities where a team can define itself, and if this could be pulled off, it would be unbelievable."*

— Dan Schulman, CEO, Virgin Mobile USA

Schulman accepted the challenge in the summer of 2001 and began to assemble a team to develop the new Virgin-branded service with a launch date of July 2002. Schulman had 18 years of telecommunications experience with AT&T and had most recently been CEO of Priceline.com. He would need to draw on his experiences from both firms to create an appealing offer that would take off in a saturated market. His goal was to achieve a run rate in which Virgin Mobile would have 1 million total subscribers by the end of the first year, and 3 million by year four.[1]

One of the key decisions for Virgin Mobile USA was the selection of a pricing strategy that would attract and retain subscribers.

## Company Background

Virgin, a U.K.-based company led by Sir Richard Branson, was one of the top three most recognized brands in Britain. The company had a history of brand extensions—more than any other major firm in the past 20 years—resulting in a vast portfolio consisting of more than 200 different corporate entities involved in everything from planes and trains to beverages and cosmetics. What tied all of these businesses together were the values of the Virgin brand:

> We believe in making a difference. In our customers' eyes, Virgin stands for value for money, quality, innovation, fun and a sense of competitive challenge. ... We look for opportunities where we can offer something better, fresher and more valuable, and we seize them. We often move into areas where the customer has traditionally received a poor deal,

---

[1] Numbers in this case are disguised for competitive reasons and utilize primary data from industry analysts.

Professor Gail McGovern prepared this case. HBS cases are developed solely as the basis for class discussion. Certain details have been disguised. Cases are not intended to serve as endorsements, sources of primary data, or illustrations of effective or ineffective management.

Copyright © 2003, 2007 President and Fellows of Harvard College. To order copies or request permission to reproduce materials, call 1-800-545-7685, write Harvard Business School Publishing, Boston, MA 02163, or go to http://www.hbsp.harvard.edu. No part of this publication may be reproduced, stored in a retrieval system, used in a spreadsheet, or transmitted in any form or by any means—electronic, mechanical, photocopying, recording, or otherwise—without the permission of Harvard Business School.

and where the competition is complacent. . . . We are pro-active and quick to act, often leaving bigger and more cumbersome organizations in our wake.[2]

Many of the company's ventures, such as Virgin Music Group, had proven to be phenomenally successful; others, such as Virgin Cola, had resulted in failure. Virgin's cellular operations in the U.K. had been among the company's success stories—Virgin had signed up approximately 2.5 million customers in just three years. The venture had broken new ground by being the country's first mobile virtual network operator (MVNO), which meant that rather than investing in and running a network in-house, the company leased network space from another firm, Deutsche Telekom.

In Singapore, however, the story had been different. There, the company's cellular service—a joint venture with Singapore Telecommunications—had run into difficulties, attracting fewer than 30,000 subscribers after its launch in October 2001. The Singapore MVNO had recently shut its doors, and although both partners had agreed that the market had been too saturated to sustain a new entrant, some analysts had offered another explanation for the failure: Virgin's hip and trendy positioning had failed to strike a chord in the Singapore market.

Despite this setback, Virgin had forged ahead with its plans to launch a wireless phone service in the U.S. Utilizing the MVNO model once again, the company had entered into a 50-50 joint venture with Sprint in which Virgin Mobile USA's services would be hosted on Sprint's PCS network. (Sprint was in the process of updating its network and increasing its capacity, so that it had ample capacity to allow for additional users.) Under the agreement, Virgin Mobile would purchase minutes from Sprint on an as-used basis.

"The nice thing about this model is that we don't have to worry about huge fixed costs or the physical infrastructure," said Schulman. "We can focus on what we do best—understanding and meeting customer needs."

## The Crowded Cellular Market: Identifying a Niche

The team leading Virgin Mobile USA was acutely aware of the overcrowded nature of the mobile communications industry in the United States. At the end of 2001, the U.S. had six national carriers and a number of regional and affiliate providers. Industry penetration was close to 50% with about 130 million subscribers, and the market was considered to have reached maturity. (Please see **Exhibit 1** for subscribers by carrier.)

Among consumers aged 15 to 29, however, penetration was significantly lower, and the growth rate among this demographic was projected to be robust for the next five years.[3] (Please see **Exhibit 2** for growth rates.)

Still, as Schulman observed, "The big players haven't targeted this segment." One reason was that young consumers often had poor credit quality. "These are people who don't necessarily have credit cards and often don't pass the credit checks that the cellular contracts require," Schulman noted.

In addition, in an industry in which the average cost to acquire a customer was roughly $370, many carriers did not believe it was worth acquiring consumers who might not use their cell phones on a frequent basis. "The assumption is that if you're not using the phone for business or if you don't

---

[2] Source: Company Web site.

[3] Source: Strategis Group.

already subscribe to a cell phone service, then you're probably not going to be someone who uses their cell phone a lot," explained Schulman. In fact, the average monthly cell phone bill for the national carriers was $52, representing about 417 minutes of use. Because the cost to serve a customer was roughly $30 a month, the carriers tended to be wary of acquiring low-value subscribers.

Despite these challenges, the Virgin Mobile team decided that this segment represented the greatest opportunity. "This is a market that has been underserved by the existing carriers," explained Schulman. "They have specific needs that haven't been met." He continued:

> A lot of the consumers in this age group are in flux in their lives. They're either in college, they're just leaving their home, or they may be getting their first cell phone. Their usage is probably inconsistent. One month, they may not use the phone at all, and another month, they may use it quite a bit, depending on whether they're on vacation or in school.

> Their calling patterns are different from the typical businessperson. They're more open to new things, like text messaging and downloading information using their phones. And they're more likely to use ring tones, faceplates, and graphics. In fact, some of them need to go to "ring tone anonymous," that's how addicted they are. Phones are more than a tool for these young people; they're a fashion accessory and a personal statement.

## VirginXtras

*The rock in our slingshot in this battle of David versus many Goliaths is focus. By focusing exclusively on the youth market from the ground up, we're putting ourselves in a position to serve these customers in a way that they've never been served before.*

—Dan Schulman

The Virgin Mobile USA team quickly began to seek ways to develop a value proposition that would appeal to the youth market. Because revenue for mobile entertainment was projected to increase steadily over the next few years (see **Exhibit 3**), the team decided that a key part of the Virgin Mobile service would involve the delivery of content, features, and entertainment, which they called "VirginXtras." To this end, the company signed an exclusive, multiyear content and marketing agreement with MTV networks to deliver music, games, and other MTV-, VH1-, and Nickelodeon-based content to Virgin Mobile subscribers. (See **Exhibit 4** for screenshots.) The deal ensured that subscribers would have access to MTV-branded accessories and phones, as well as branded content such as graphics, ring tones, text alerts, and voice mail. The company would also receive promotional airtime on MTV's channels and Web site. And under the agreement, Virgin Mobile subscribers would be able to use their phones to vote for their favorite videos on shows like MTV's "Total Request Live." As Schulman put it:

> We're taking cell phone content to a whole new level. It's a great match: MTV Networks is home to some of the most recognized youth brands in the country; it has unparalleled reach for the under-30 market. The Virgin brand is all about fun, honesty, and great value for money, which is what our target market wants. You put the two together, and you've got some of the most exciting cell phone features in the market. It's a powerful relationship for us.

In addition to the MTV-branded content, the Virgin Mobile service would also include the following VirginXtras:

- **Text Messaging.** Schulman believed text messaging was a key selling point for youth: "The number of text messages tends to skyrocket during school hours. Kids discreetly text message while they're in class. Part of the reason why they communicate like this is so their parents don't see who they call. It's a very private form of communication for them."

- **Online Real-Time Billing.** For additional privacy from their parents, kids would not have call detail on monthly bills. Virgin Mobile would provide a Web site with a record of individual calls on a real-time basis.

- **Rescue Ring.** Virgin Mobile subscribers would be able to schedule a "rescue ring," which would call them at a prearranged time to provide them with an "escape" in case a date was not going well. If the date was going well, they could always tell the "caller" that they would get back to them tomorrow.

- **Wake-Up Call.** For those who needed a little help getting out of bed in the morning, Virgin Mobile USA would offer its customers the chance to wake up to original messages from a variety of cheeky celebrity personalities.

- **Ring Tones.** A large selection of tunes would be available for subscribers to download if they wanted to customize their ring tones, ranging from hip hop to rock to the Sponge Bob Square Pants anthem.

- **Fun Clips.** These audio clips would consist of news tidbits, jokes, gossip, sports information, and more.

- **The Hit List.** Subscribers would be able to use their handsets to listen to and vote on a top 10 list of hit songs. After voting, customers would be able to hear the percentage of other subscribers who either "loved it" or "hated it."

- **Music Messenger.** This service would let subscribers tap into a top 10 song list and then would shoot a message to a friend allowing them to check out a hot new track.

- **Movies.** This service would provide movie descriptions, show times, and allow subscribers to buy tickets in advance using their phones.

The Virgin team believed that these features would appeal to the youth market, generate additional usage, and create loyalty. Schulman elaborated: "Our market research indicates that VirginXtras will attract and retain the youth segment. Not only will these features be appealing, but we also believe they will be addictive and will bond our customers to their cell phones."

## Purchasing the Service

Most cellular providers sold their services in their own proprietary retail outlets, kiosks in malls, high-end electronic stores (e.g., Radio Shack), specialty stores, and so on. Because these retail outlets typically employed high-touch salespeople, most providers paid high sales commissions to ensure hands-on service.

In contrast, the Virgin Mobile team had already decided to adopt a different channel strategy that was more closely aligned to its target-market selection. Schulman explained:

We've decided to distribute in channels where youth shop. This means places like Target, Sam Goody music stores, and Best Buy. In these stores, kids are used to buying consumer electronics products. They're used to buying a CD player or an MP3 player. So we've decided to package our products in consumer electronics packaging. Instead of being in a box locked behind some counter, we've created a clamshell, clear, see-through package where consumers can pick up the phone without a salesperson helping them and purchase it like they would any other consumer electronics product.

Cellular carriers historically purchased handsets from cell phone manufacturers such as Nokia, Motorola, Samsung, and Lucky Goldstar. Although the cost per handset generally ranged from $150 to $300, carriers typically charged end users between $60 and $90.[4] This handset subsidy was an accepted part of the carrier's acquisition costs.

Virgin had a contract with handset manufacturer Kyocera by which it would buy phones for anywhere from $60 to $100 depending on the features and functions of the phones.[5] The first two basic models would be named the "Party Animal" (a Kyocera 2119) and the "Super Model" (a Kyocera 2255). Both would come bundled with interchangeable faceplates that would be decorated with eye-catching colors and patterns (see **Exhibit 5** for sample phones) and would be nestled inside one of Virgin Mobile's bright red clamshell-style Starter Packs (see **Exhibit 6** for pictures of packaging).

The Starter Packs would be easily visible on large point-of-sale displays (see **Exhibit 7**) that the company would make available to its retailers. The company had entered into distribution agreements with Target and Best Buy, both of which charged lower commissions than traditional industry channels—$30 per phone, versus an industry average of $100.[6] The Starter Packs would also be available at retailers such as Sam Goody, Circuit City, Media Play, and Virgin Megastores. In total, the company expected its phones to be available at more than 3,000 U.S. retail outlets by the time the service launched in July.

## Advertising

*Unless you're between 14 and 24, you're probably never going to see our ads. If you ever see us on "60 Minutes," then you know we've gone astray. Think WB, MTV, and Comedy Central [three youth-oriented networks].*

— Dan Schulman

The U.S. cellular industry was projected to spend about $1.8 billion in advertising in 2002. Most national carriers had huge ad budgets; for example, Verizon Wireless alone was expected to spend more than $650 million advertising in major media in 2002.[7] Virgin Mobile USA's advertising budget was miniscule by comparison: approximately $60 million.

Still, Schulman was determined to make the most of the limited budget. "By definition, the big players need to be all things to all people. They are throwing huge amounts of money into messages

---

[4] Source: Morgan Stanley research.

[5] Numbers are disguised for competitive reasons.

[6] Numbers are disguised for competitive reasons.

[7] Source: TNS Media Intelligence/CMR. For the national carriers, advertising spending typically ranged from $75 to $105 per customer acquired.

that are largely undifferentiated," he said. "Our goal is different; we want to break through the clutter. Our advantage is that we've got a much tighter focus on a much narrower target market; this means we have to be able to get our message out more efficiently than our competitors."

The team had already decided on an advertising campaign that it believed was quirky, offbeat, and completely different from competitive ad treatments. The ads would feature teens and would make use of strange, often-indecipherable metaphors. As Howard Handler, Virgin Mobile's Chief Marketing Officer, put it, "We need to stand out from the rest of the crowd, which means that we need to deliver ads that are not run-of-the-mill. They need to be more entertaining and more unique in their creative execution." In addition, the company was working with youth magazine editors of publications such as *The Complex*, *Vibe*, and *XXL* to publish "advertorials," pieces extolling Virgin Mobile to their readers. "These are the opinion-leading magazines," Handler said. "Getting their buy-in is important for us."

Virgin Mobile was also planning a number of high-profile street marketing events. These events would feature paid performers—dancers and gymnasts dressed in red from head to toe—who would engage in various stunts.

Finally, the team was in the process of planning a highly unusual event to kick off the launch of the Virgin Mobile USA service. The plan called for the cast of *The Full Monty*, a Broadway show, to appear with Sir Richard Branson, dangling from a building in New York City's Time Square, wearing nothing but a large, strategically placed cell phone. (See **Exhibit 8** for pictures from the launch.)

## The Pricing Decision

*We knew that we couldn't afford to get pricing wrong when we designed our offer. It can make or break your success. Consequently, we did a tremendous amount of market research among our target segment, and one thing became clear: Our audience did not trust the industry pricing plans. They all advertise "free this" and "free that," but young people know that there are a lot of hidden charges, and they resent this. These are savvy consumers, and they hate feeling like they're being conned. So we've got an opportunity to use pricing as a way to differentiate ourselves from the competition.*

—Dan Schulman

Over 90% of all subscribers in the U.S. had contractual agreements with their cellular providers. The contracts were generally for a period of one to two years, and they required a rigorous credit check. Many plans had established "buckets" of minutes. Customers could sign up for a bucket of 300 minutes, for example. However, if they actually used more than 300 minutes, they were penalized with extremely high rates (e.g., 40 cents/minute) for the overage. If they used fewer than 300 minutes, they were still charged the fixed monthly fee, which then drove up their price per minute.

The carriers typically charged less for off-peak than on-peak minutes, but the off-peak period had shrunk over time. Originally, off-peak had begun at 6:00 p.m.; the starting time had since shifted to 7:00 p.m., then 8:00 p.m., and finally 9:00 p.m. Some carriers such as Cingular charged a monthly fee (about $7) to move the peak time back one hour. Schulman noted:

> The industry is making money from customer confusion. As a customer, you need to use minutes within the tight range that you signed up for in order to get a good rate. Your on-peak and off-peak minutes have to be in the right mix too. If all customers actually signed up for the optimal plan for their usage, the carriers would be making far less money than they are today.

In fact, the industry's pricing plans were quite rational if customers would always select the right plan for their usage patterns. (Please refer to **Exhibit 9a**.) However, customers usually could not predict their usage. Virgin Mobile studied hundreds of customers and found that the prices they actually paid varied widely. (Please refer to **Exhibit 9b**.) Schulman continued:

> Often customers *think* that they use more minutes than they *actually* use. For example, in our target segment, the majority of young people actually use from 100 to 300 minutes per month. However, if you ask them to predict their usage, they'll often come up with a much higher number. Other people will try to pick lower bucket plans to avoid high monthly fees. Then they'll get a $100 bill because they didn't realize that it would cost them 40 cents for every minute above the bucket.

Adding to consumer resentment was the fact that most carriers slapped on additional fees to add to the monthly bill. Schulman explained: "The carriers will only tell you about the monthly bucket fee; they won't mention the taxes you'll have to pay or the universal services charge that you'll have to pay. There are a bunch of one-time costs that are loaded on top of the bill that they don't advertise. So even if you end up being exactly right in your bucket, a $29 plan ends up being a $35 plan."

Schulman and his team carefully considered various pricing strategies. Although the pricing possibilities were endless, the team believed that there were realistically three viable options. Schulman said: "We're trying to be as open-minded as possible. We have the luxury of starting from scratch, so this is an opportunity to fix some of the problems that are endemic in this industry. Our only constraints are that (1) we want to make sure our prices are competitive, (2) we want to make sure we can make money, and (3) we don't want to trigger off competitive reactions."

## *Option 1—"Clone the Industry Prices"*

The first option was to merely "clone" the existing industry price structure. (See **Exhibit 10a** for Option 1 pricing.) All of the major carriers paid high commissions to salespeople to explain their complicated pricing structures and to perform credit checks. (In fact, 30% of prospective customers failed to pass these credit checks.) Given Virgin Mobile's nontraditional channel strategy, its pricing message would have to be relatively simple. Schulman said, "With this first option, we would simply be telling consumers that we're priced competitively with everyone else, but with a few key advantages like differentiated applications [MTV] and superior customer service."

In addition, Virgin Mobile could attempt to differentiate from the competition by offering better off-peak hours and fewer hidden fees. "We know that consumers are sick of hidden fees and they hate off-peak deals that start at 9 p.m., so we'd be addressing a real sore spot among young people," said Schulman.

He added, "The nice thing about this idea is that it's easy to promote. People may not like the pricing plans, but given all the money the industry spends to promote them, the customers are used to 'buckets' and peak/off-peak distinctions. Given our limited advertising budget, it may be a stretch for us to break through with anything different. We could also put it on our packaging so that even without the help of a salesperson, consumers would get the message."

## *Option 2—"Price Below the Competition"*

The second option was to adopt a similar pricing *structure* as that of the rest of the industry, with *actual* prices slightly below those of the competition. That is, Virgin Mobile would maintain the buckets and volume discounts, but its price per minute would be set below the industry average for certain key buckets (see **Exhibit 10b**).

"This option would allow us to tell consumers that we're cheaper, plain and simple. Because our target market generally uses between 100 and 300 minutes per month, that's where consumers would get the best price," said Schulman. "Under this option, we could also offer better off-peak hours and fewer hidden fees, but I don't know if that would be necessary if our price per minute was clearly below the competition. We wouldn't want to leave too much money on the table."

*Option 3—"A Whole New Plan"*

The third option was the most radical. The idea was to start from scratch and come up with an entirely different pricing structure, one that was significantly different from anything offered by the competition. The pricing variables that Schulman was toying with included:

- **The role of contracts.** Did it make sense to shorten the term of the subscription contracts, or perhaps even eliminate the contracts altogether? Contracts provided carriers with a hedge against churn and a guaranteed annuity stream; yet even with the contracts, cellular providers struggled with an industry churn rate that averaged 2% per month. If Virgin Mobile were to shorten or eliminate such contracts, the risk would be that its churn rate would skyrocket. In fact it was estimated that churn would climb to 6% each month.[8]

Schulman added:

> From a marketing perspective, there's no question that it would be great if we could announce to the world that we've eliminated contracts. Keep in mind that, if you're under 18, you can't even enter into a contract with a cellular provider. Your parents need to do it for you. So eliminating contracts would be a big advantage for us from a customer-acquisition standpoint. Of course, in terms of retention, contracts are a safety net. So the question is, does it make sense for us to try to fly without a safety net?

- **Prepaid versus post-paid.** The vast majority (92%) of current cell phone subscribers in the U.S. had post-paid plans, which meant that they were billed monthly on the basis of their contract. Prepaid arrangements, in which consumers purchased a number of minutes in advance, were unusual because of prohibitive pricing (generally, between 35 and 50 cents per minute, and as high as 75 cents per minute). Most prepaid customers used their phones on an occasional basis as a safety device: "They just keep them in their glove compartment," as Schulman put it. Many of these customers had poor credit; in fact, the reason prepaid plans appealed to them was that such plans required no credit checks. Customers therefore thought that prepaid arrangements were a stigma, and the prepaid offers tended to attract low-usage customers. Still, in countries such as Finland and the U.K., prepaid arrangements were commonplace, accounting for the majority of new gross adds.

Schulman knew that the risks of adopting a prepaid pricing structure were significant. U.S. carriers were extremely wary of prepaying consumers because of their high churn rates; prepaying consumers tended to exhibit no loyalty to a provider once they had used up all of their prepaid minutes. If Virgin Mobile were to adopt a prepaid pricing structure, the danger was that the company would never be able to recoup its customer acquisition costs. In fact,

---

[8] Source: Morgan Stanley research.

industry analysts estimate that total acquisition costs would have to be at or below $100 per new gross add for prepaid to be viable.[9]

In addition, there were a number of related issues to consider. A prepaid pricing structure would require some mechanism—perhaps via the Web or through physical phone cards—whereby consumers could easily add minutes to their phone.

- **Handset subsidies**. Most carriers purchased handsets from cell phone manufacturers such as Nokia, Motorola, and Samsung at a cost per handset ranging from $150 to $300 for the industry. The carriers then subsidized the cost of the handset to end users. This subsidy—which was typically about $100 to $200—was part of the customer acquisition cost.

  "We're debating all of our options here," said Schulman, "everything from increasing the subsidy so that our phones are cheaper than the competition, to lowering the subsidy as a way of getting consumers to feel more invested and loyal towards our service."

- **Hidden fees and off-peak hours.** "One of our goals is to offer a service that is priced so simply that consumers don't need a math degree to figure it out," noted Schulman. "One way to do this would be to eliminate *all* hidden fees, including taxes, universal service charges, *everything*. It would literally be 'what you see is what you get.' However, this would mean rolling all of those hidden costs into our pricing structure in such a way that our pricing feels competitive to our target market, and yet we still make money."

  As for off-peak hours, "We need to think about what makes sense for our target customer," said Schulman. "These kids don't lead the same kind of lifestyle as the typical business-person, so our service should define off-peak with that in mind."

As Schulman reviewed the various options for pricing, he realized the importance of laying the foundation for future profitability. "There's this assumption that you can't target young people and make money," he said. "Our goal is to prove otherwise. Ideally, every customer we acquire will have positive lifetime value (LTV) for us." (See **Exhibit 11** for LTV details.)

"That's why this pricing decision is so critical," he continued. "If we can figure out a way to create value so that we can successfully enter a very competitive and saturated market, and also create profitability with this target segment, then we will have truly accomplished something big."

---

[9] Source: Morgan Stanley research.

**Exhibit 1** Wireless Subscribers in the United States, by Carrier (Q4 2001, in millions)

| Carrier | Subscribers |
| --- | --- |
| AT&T (affiliates) | 20.5 |
| Cingular | 21.7 |
| Verizon | 29.5 |
| VoiceStream | 6.5 |
| Alltel | 6.7 |
| Sprint | 14.5 |
| U.S. Cellular | 3.5 |
| Leap | 1.1 |
| Other Carriers | 26.1 |
| Total | 130.0 |

Source: Adapted from The Yankee Group.

**Exhibit 2** Mobile Penetration by Age Group

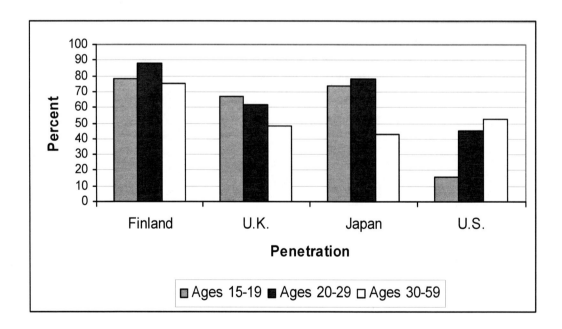

Source: Adapted from IDC, Salomon Smith Barney.

**Exhibit 3**  Revenue from Mobile Entertainment Services

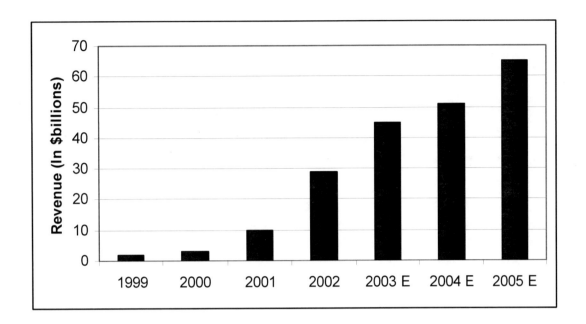

Source: Adapted from The Yankee Group.

Note: Revenues include video, audio, graphics, and games.

**Exhibit 4**   Screenshots of Virgin Mobile USA Content

Source: Company Web site.

**Exhibit 5**   Virgin Mobile USA Handset Models

Source:   Company Web site.

Note:   Phones in second row show various faceplates for a single model.

**Exhibit 6** Virgin Mobile USA: The Super Model Starter Pack (clamshell packaging)

Source: Company Web site.

Virgin Mobile USA: Pricing for the Very First Time

**Exhibit 7**   Virgin Mobile USA Point-of-Sale Displays

Source:   Company Web site.

**Exhibit 8** Picture of Branson at Launch

Source: *Forbes* Magazine, October 7, 2002.

©Lawrence Lucier/Getty Images

Virgin Mobile USA: Pricing for the Very First Time

**Exhibit 9a** Calling Plans—Industry Prices

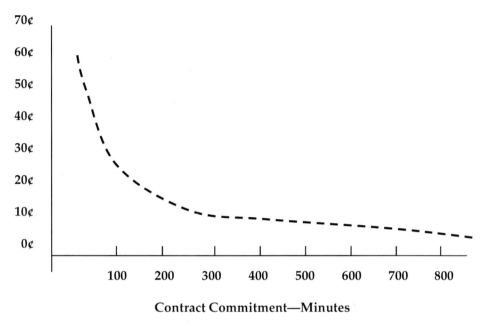

Source: Adapted from company data, Morgan Stanley research.

**Exhibit 9b** Actual Prices Paid by Customers

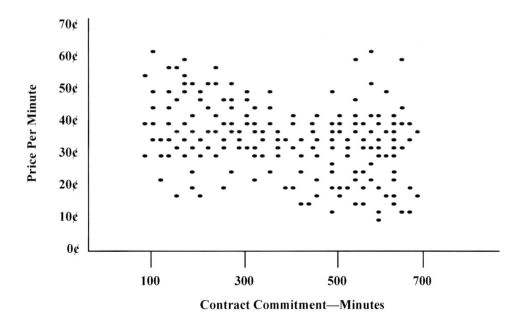

Source: Adapted from company data, Morgan Stanley research.

**Exhibit 10a** Option 1 Pricing Structure

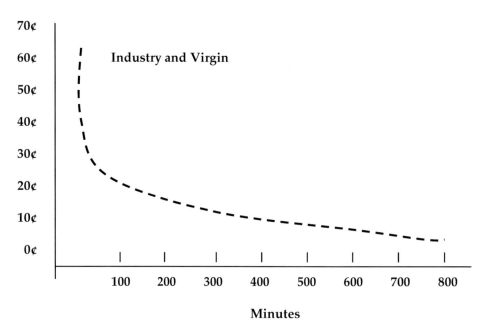

Source: Adapted from company data, Morgan Stanley research.

**Exhibit 10b** Option 2 Pricing Structure

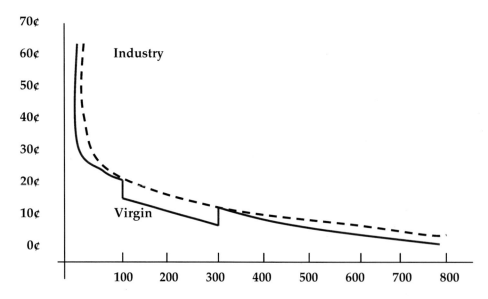

Source: Adapted from company data, Morgan Stanley research.

Note: Prices are for a blend of on- and off-peak minutes, with off-peak beginning at 9:00 p.m. Each additional off-peak hour reduces average price per minute by approximately 1.5 cents.

**Exhibit 11** Calculating Lifetime Value (LTV) for Cellular Subscribers

In general, lifetime value (LTV) for a customer is calculated as follows:

$$LTV = \sum_{a=1}^{N} \frac{(M_a) r^{(a-1)}}{(1+i)^a} - AC$$

where

$N$ = the number of years over which the relationship is calculated

$M_a$ = the margin the customer generates in year a

$r$ = the retention rate ($r^{(a-1)}$ is the survival rate for year a)

$i$ = the interest rate

$AC$ = the acquisition cost

Source: Adapted from "Customer Profitability and Lifetime Value," HBS Note 503-019.

In the cellular industry, margin is relatively fixed across periods. Therefore, one can simplify the above expression by assuming an infinite economic life (i.e., letting $N \rightarrow \infty$), which leads to:

$$LTV = \frac{M}{1-r+i} - AC$$

*Monthly Margin = average revenue per unit per month (ARPU) – monthly cost to serve (CCPU, or cash cost per user)*

The components of AC were advertising per gross add, the sales commission paid per subscriber, and the handset subsidy provided to the subscriber.

CCPU consisted of customer-care costs, network costs (the cost of using Sprint's network), IT costs, and overhead. Industry analysts estimated that Virgin Mobile's CCPU would be constant at 45% of revenues during its first year of operations, since most of Virgin's costs were variable. Monthly churn was estimated to be 2% for customers under contract and 6% for prepaid customers.[a]

Interest rates were 5%.

[a]Numbers disguised for competitive reasons.

# Harvard Business School

9-191-058
October 19, 1990

# Beauregard Textile Company

In September 1990, Joel Calloway and Clarence Beal, Jr., sales manager and controller respectively of the Beauregard Textile Company, met to prepare recommendations for fourth-quarter fabric prices. When approved by the executive committee, fabric prices were published and mailed to customers. These prices were considered firm for the quarter. Beauregard was one of the largest firms in its segment of the textile industry with annual sales of about $82 million. Company sales persons were paid straight salaries and sold the full line of fabrics.

On this occasion, Calloway and Beal were particularly concerned on how to price Triaxx-30, a blend of nylon, polypropelene, and rayon used for special outdoor applications. In January 1990, Beauregard had raised its price for Triaxx-30 from $3 to $4 a yard in order to bring its profit margin up to that for other products. This action reflected in part an adjustment to recent increases in costs. It was also motivated by a directive from the board of directors that urged management to strengthen the company's working capital position so as to ensure the availability of adequate funds for a recently approved long-term plant modernization and expansion program. (Triaxx-30 was woven on special equipment that could not be used for other purposes.)

Calhoun & Pritchard Inc., the only significant alternative supplier of T-30 (the common trade reference for Triaxx-30-type fabric), had held its price at $3. (Calhoun & Pritchard normally waited for Beauregard to announce fabric prices before mailing its own price list.) As a result, Beauregard had lost market share. As the sales history in *Exhibit 1* shows, the total market for T-30 fabric had remained remarkably stable for the last three years at about 225,000 yards perquarter. These data also showed some customers to be price-sensitive, switching immediately to the low cost supplier. Based on conversations with his sales people, Calloway believed that a number of customers would discontinue or restrict their use of T-30 fabric if it were no longer available at $3, reducing the overall demand by 20%.

The following excerpts from the discussion reveal some of the two men's understanding and concerns about the Triaxx-30 pricing decision:

**Calloway:** *If we reduce our price to $3, ol' C and P might drop its price, leaving us worse off.*

**Beal:** *I doubt they would go below $3. For one thing, they haven't done so in the past. For another, their costs are comparable to ours. (See Exhibit 2 for the Triaxx-30 cost schedule.) And from what I hear, Calhoun & Pritchard is in a tight financial situation as a result of its recent takeover defense. What I've never been*

---

*This case was prepared by Professor Francis J. Aguilar as a basis for class discussion rather than to illustrate either effective or ineffective handling of an administrative situation. This case incorporates parts of the Atherton Company case (156-002).*

Copyright © 1990 by the President and Fellows of Harvard College. To order copies or request permission to reproduce materials, call 1-800-545-7685 or write Harvard Business School Publishing, Boston, MA 02163. No part of this publication may be reproduced, stored in a retrieval system, used in a spreadsheet, or transmitted in any form or by any means—electronic, mechanical, photocopying, recording, or otherwise—without the permission of Harvard Business School.

able to figure out is why they didn't move up to $4 when we went up. They've got to be losing money at $3. It just doesn't make sense.

**Calloway:** Well, Clarence, that's their problem. Our problem is if we both continue to charge current prices, our order book remains depressed. My people would prefer to drop the price to $3 and regain our old customers. They'll come back with price parity because of our location advantage. And dropping the price from $4 to $3 won't affect the sales of any of our other products.

**Beal:** That may be well and good for sales morale, but how do we justify pricing below cost?

**Calloway:** This choice between $3 and $4 is beginning to look more difficult than I thought. So which is it to be?

**Exhibit 1** Quarterly Prices and Sales Volumes for T-30 Fabric, 1988-1990

| Year and Quarter | | Beauregard | | Calhoun & Pritchard | |
|---|---|---|---|---|---|
| | | Price | Actual Sales Volume (Yards) | Price | Estimated Sales Volume (Yards) |
| 1988 | 1st | $3 | 124,870 | $3 | 100,000 |
| | 2nd | 3 | 126,016 | 3 | 100,000 |
| | 3rd | 3 | 125,426 | 3 | 100,000 |
| | 4th | 3 | 198,863 | 4 | 25,000 |
| 1989 | 1st | 3 | 127,201 | 3 | 100,000 |
| | 2nd | 3 | 125,277 | 3 | 100,000 |
| | 3rd | 3 | 126,124 | 3 | 100,000 |
| | 4th | 3 | 125,302 | 3 | 100,000 |
| 1990 | 1st | 4 | 74,860 | 3 | 150,000 |
| | 2nd | 4 | 77,216 | 3 | 150,000 |
| | 3rd | 4 | 75,000 (est.) | 3 | 150,000 |
| | 4th | | | | |

Source: Marketing Department, Beauregard Textile Company

**Exhibit 2** Beauregard's Estimated Cost per Yard of Triaxx-30 at Various Volumes of Production

|  | Production Volume in Yards of Material | | | | | | | |
|---|---|---|---|---|---|---|---|---|
|  | 25,000 | 50,000 | 75,000 | 100,000 | 125,000 | 150,000 | 175,000 | 200,000 |
| Direct Labor | $ .860 | $ .830 | $ .800 | $ .780 | $ .760 | $ .740 | $ .760 | $ .800 |
| Material | .400 | .400 | .400 | .400 | .400 | .400 | .400 | .400 |
| Material Spoilage | .042 | .040 | .040 | .040 | .038 | .038 | .038 | .040 |
| Department Expense: | | | | | | | | |
|   Direct* | .198 | .140 | .120 | .112 | .100 | .100 | .100 | .100 |
|   Indirect** | 2.400 | 1.200 | .800 | .600 | .480 | .400 | .343 | .300 |
| General Overhead*** | .258 | .249 | .240 | .234 | .228 | .222 | .228 | .240 |
| Factory Cost | $4.158 | $2.859 | $2.400 | $2.166 | $2.006 | $1.900 | $1.869 | $1.880 |
| Selling and Admin. | | | | | | | | |
|   Expense**** | 2.703 | 1.858 | 1.560 | 1.408 | 1.304 | 1.236 | 1.215 | 1.222 |
| Total Cost | $6.861 | $4.717 | $3.960 | $3.574 | $3.310 | $3.136 | $3.084 | $3.102 |

    \*   Indirect labor, supplies, repairs, power, etc.
   \*\*   Depreciation, supervision, etc.
  \*\*\*   30 percent of direct labor
 \*\*\*\*   65 percent of factory cost

Richard Ivey School of Business
The University of Western Ontario

9B07A004

# TUTTI MATTI

*Ian Da Silva prepared this case under the supervision of Elizabeth M. A. Grasby solely to provide material for class discussion. The authors do not intend to illustrate either effective or ineffective handling of a managerial situation. The authors may have disguised certain names and other identifying information to protect confidentiality.*

*Ivey Management Services prohibits any form of reproduction, storage or transmittal without its written permission. This material is not covered under authorization from CanCopy or any reproduction rights organization. To order copies or request permission to reproduce materials, contact Ivey Publishing, Ivey Management Services, c/o Richard Ivey School of Business, The University of Western Ontario, London, Ontario, Canada, N6A 3K7; phone (519) 661-3208; fax (519) 661-3882; e-mail cases@ivey.uwo.ca.*

Copyright © 2006, Ivey Management Services                    Version: (A) 2007-01-19

On May 1, 2003, Alida Solomon turned off the evening news in frustration after another discouraging report on the economic effects of the Severe Acute Respiratory Syndrome (SARS) outbreak in Toronto, Canada. In November 2002, in Toronto, Solomon had just opened her first restaurant, Tutti Matti, where she was the head chef. Offering a variety of authentic meals and wine from the Italian region of Tuscany, the restaurant had begun to attract a growing customer base. Although it still lacked the consistent business and media acclaim vital to the success of an up-and-coming Toronto restaurant, Solomon was encouraged by the restaurant's early performance. The first year of operations was critical to a restaurant's long-term survival. Solomon knew that to continue the restaurant's success and to overcome the added challenges created by SARS, she would need a strong and creative marketing plan.

## ALIDA SOLOMON

Alida Solomon's first experience as a chef came as a teenager, and she knew immediately that she had found her calling in life. After high school, under her parents' instruction, Solomon enrolled in the General Arts program at Montreal's Concordia University. Early in her first year, however, Solomon realized that university studies did not interest her so, the following September, she enrolled in one of Canada's leading culinary management programs at Toronto's George Brown College. Solomon excelled at George Brown, and at the end of her two-year program, she graduated at the top of her class. Throughout her studies, Solomon was employed as a pastry chef and kitchen understudy at a popular Toronto restaurant, Galileo, where she gained valuable hands-on experience and dreamed of one day becoming the head chef of her own restaurant.

Wanting to improve her skills and unhappy with the opportunities in Toronto, Solomon packed up her belongings upon graduation and moved to the Italian region of Tuscany, world famous for its red wine and hearty cuisine. It was here that Solomon believed she could excel as an aspiring chef. She settled in Montalcino, a popular tourist town of 5,000 residents, which boasted nearly 200 active wineries. (See Exhibit 1 for a map of Tuscany and Italy.) Within a few months, Solomon had secured a position as an apprentice chef at one of the region's most respected authentic Italian restaurants, along with a job picking grapes at a local vineyard.

Solomon spent the next six years in Tuscany learning all aspects of a restaurant's operations. She did everything from butchering meat and selecting farm-fresh vegetables to preparing all meals. Everything was done in the traditional way — work was completed as a family, and there was no such thing as a formal job description. She loved the team atmosphere and, although she found work in the Tuscan kitchen challenging, it was very exciting.

## TUTTI MATTI

### History

It was during her time in Tuscany that Solomon's dream for Tutti Matti was conceived. She adored the Italian lifestyle, which she described as warm, relaxed and family-oriented, and she knew that she would love to own a restaurant reflecting that lifestyle. Solomon observed that many of the visitors to Tuscany were Canadian, and most raved about Tuscan cuisine. Solomon recalled countless Italian restaurants operating in Toronto, but none of them were regionally focused, Tuscan restaurants. With this knowledge, Solomon suspected that there could be a very attractive opportunity for her to bring home her skills and passion for Tuscan cuisine. Solomon did not have much entrepreneurial experience, but she knew that opening a restaurant required a hefty financial investment and that it could also be very risky.[1]

Nonetheless, Solomon was very excited and she quickly grew determined to make her dream a reality. While still working in Tuscany, Solomon enlisted the help of her father, a lawyer with experience in commercial real estate. Solomon's father

---

[1] According to Forbes magazine, opening a new restaurant costs at least US$185,000, and one in two new restaurants fails within its first two years in business. Forbes Magazine, "No Free Lunch," http://www.forbes.com/free_forbes/2003/0609/154.html, November 15, 2005.

bought in to his daughter's idea and, while Solomon spent her last year in Tuscany, her father began to search for a location in Toronto. After finding a suitable location, Solomon and her father secured a small business loan with a major Canadian bank, and Solomon returned to Toronto to pursue her dream.

## SARS OUTBREAK

Solomon returned from Italy in early 2001 and, after 18 months of hard work, Tutti Matti opened its doors in November 2002. At around the same time as Tutti Matti was getting off the ground, thousands of miles away, in China, there was an outbreak of the Severe Acute Respiratory Syndrome (SARS) virus. Eventually, the virus made its way to Toronto where the first Canadian SARS-related death occurred in Toronto in early March 2003. The presence of SARS in Canada became a feature on news broadcasts worldwide.

SARS is a viral respiratory illness that was first reported in Asia in early 2003. Of a total of 8,098 people infected, 774 eventually died from the virus. The spread of the virus comes mainly from close personal contact, but airborne transmission is also possible, thereby increasing the risk of infection. Health officials believe that the most ready transmission comes from respiratory droplets that are produced during a cough or a sneeze.[2]

Outside Asia, Toronto was one of the cities most affected by the virus, which is believed to have travelled to Toronto via an infected traveller. The virus could infect anyone in contact with it, but one of the major concerns in Toronto was for hospital workers who were at the highest risk of infection. Various measures, including reduced visiting privileges, stringent hand sanitization and respiratory mask programs, were implemented at all Toronto hospitals to decrease the risk of transmission.

By April 2003, Canadian health officials reported nine deaths caused by the virus and over 130 suspected and probable cases outstanding. On April 23, the World Health Organization issued a warning against all but essential travel to Toronto.[3] The city's tourism and hospitality industries encountered devastating decreases in tourism spending due to the negative media attention. Toronto's restaurants and hotels scrambled to increase business, and it was reported that the city had lost as much as $39 million in accommodations revenue alone during the month of April.[4]

---

[2] Centers for Disease Control and Prevention, http://www.cdc.gov/ncidod/sars/factsheet.htm, August 25, 2005.
[3] CBC News, "INDEPTH: SARS, Timeline," http://www.cbc.ca/news/background/sars/timeline.html, August 25, 2005.
[4] Canada Tourism.com, http://www.canadatourism.com/ctx/app/en/ca/pressItem.do?articleId=46842&language=english, August 25, 2005.

## TUTTI MATTI

### The Restaurant's Name

Solomon wanted her restaurant's name to reflect its authentic Italian offering and to convey a specific atmosphere. She believed that Tutti Matti, meaning "everybody is crazy" in Italian, was the perfect name — not only was it Italian, but it also reflected the fun-loving attitude that she wanted guests to remember as part of the Tutti Matti dining experience. While Tutti Matti would offer a first-class menu, Solomon wanted her restaurant's service and atmosphere to be distinguished by a light-hearted twist — one of the characteristics that Solomon herself was known for.

### Location

Many critics had referred to Tutti Matti's location as "a little off the beaten path," near King Street and Spadina Avenue (see Exhibit 2) in downtown Toronto. The area was not a popular destination in itself, but it was a short walk in each direction from a heavy-traffic district. Chinatown and the trendy shopping of Queen Street West were to the north, the theatre and entertainment districts to the east, a budding bar and nightclub district to the west, and Toronto's professional sports complexes — the Rogers Centre (Skydome) and the Air Canada Centre — to the south. With heavy condominium construction in the area, King and Spadina was becoming one of the popular areas in which to live in the downtown core.

Solomon selected Tutti Matti's location because of the potential she saw for transforming the building into her own unique space. One of Tutti Matti's distinctive features was its open-concept kitchen. The kitchen was located in the middle of a 70-seat dining room and was completely open, allowing diners to watch Solomon and her team at work (see Exhibit 3). Customers enjoyed the open-kitchen experience, and they appreciated the aromas and sounds of a working kitchen; at times, customers left their seats to watch the cooking. When speaking to her clients, Solomon discovered that the kitchen was a popular feature because customers enjoyed the opportunity to watch the meal preparation process, without actually having to work.

By leasing the building[5] and purchasing the majority of her equipment from a restaurant bankruptcy sale, Solomon had minimized her initial investment. In total, her start-up costs were almost $380,000, which included $100,000 spent on equipment (e.g. appliances and dining wares), and $280,000 spent on construction and building upgrades. No major purchases were anticipated for the next year.

---

[5] *Solomon had negotiated a rate of $6,000 a month for the lease.*

With sales during the first six months of $210,000,[6] Solomon hoped to finish the year at a minimum of $500,000 in gross sales.

**The Menu**

Tutti Matti offered authentic, regional Tuscan cuisine. To ensure that her menu remained true to its Tuscan roots, Solomon prepared all food from scratch. Everything from daily-baked bread to pastas and sauces were created in Tutti Matti's kitchen, based on Solomon's own Tuscan recipes. Only the highest quality ingredients were used. All meat was purchased from a premium butcher, and Solomon personally selected all produce directly from distributors at the Toronto food terminal. Since Solomon bought the majority of ingredients herself, the step of using a food distributor was eliminated. She estimated that this saved nearly $15,000 annually. While this practice was very labor intensive, Solomon thought it was necessary to providing the superior quality promised by the restaurant.[7]

Every dish was listed on the menu by its Italian name, followed by a brief description of the plate in English (see Exhibit 4). Solomon had carefully selected wines to complement the menu. The extensive wine list included exclusively Italian wines, with a heavy bias toward Tuscan red wines. She prided herself on having one of Toronto's finest selections of the popular Brunello variety of red wines. Despite the high quality and rarity of many of the wines offered, Solomon wanted the wines to remain accessible to all customers, so a below average markup of 120 per cent was applied on all wine sold by the bottle.[8]

**TUTTI MATTI'S CUSTOMERS**

**The Lunch Crowd**

Tutti Matti served lunch starting at 11 a.m., Monday to Friday and, in addition to a regular menu, offered a $15 *prix-fixe*[9] option that was a popular draw for the lunchtime crowd. The lunch seating ended when patrons stopped coming, which was around 4 p.m. Most lunch customers were young professionals who worked at one of the several professional service firms that were located within a 10-minute walk of Tutti Matti. A large percentage of these customers were repeat diners who frequented the establishment at least twice a week. Lunchtime diners were interested in a reasonably priced alternative to traditional fast food options, and most were on a tight schedule. Most customers enjoyed their food at the

---

[6] *Solomon estimated that 70 per cent of revenue came from food sales, while 30 per cent was from beverages.*
[7] *Typical food costs for this industry ranged from 27 per cent to 32 per cent of sales. Tutti Matti's food costs averaged 30 per cent of sales.*
[8] *Typical markup on wine was at least 150 per cent and could reach up to 300 per cent.*
[9] *Prix-fixe is a menu option at a set price where diners create their own three-course meal by choosing one of a number of options in each of three categories – appetizer, main course and dessert.*

restaurant, but some also either called ahead of time or walked in to order and took their food to go. The average customer bill at lunchtime was $15 before tax and tip, and only about 15 per cent of lunchtime diners ordered an alcoholic drink with their meal. To date, lunchtime revenues accounted for the majority of Tutti Matti's customers and nearly 80 per cent of its total revenue.

**The Dinner Crowd**

Tutti Matti was open for dinner from 6 p.m. until 10 p.m. on weeknights, and from 6 p.m. to 10:30 p.m. on Friday and Saturday evenings.[10] The restaurant was not open on Sundays since this was Solomon's time to spend with family and friends. Dinner patrons differed from the lunch crowd: most enjoyed an appetizer and a main course, and almost 90 per cent ordered an alcoholic drink, typically wine, with their meal. Most dinner customers were on a more relaxed schedule than lunchtime diners, while some were on their way to a movie or to one of the many theatres located in the nearby Theatre District. The average bill per dinner customer was $50 plus tax and tip.

**Private Functions**

Occasionally, Tutti Matti hosted a corporate or private gathering that occupied the whole restaurant. In order to reserve the restaurant, the host had to guarantee the same level of sales that would typically be expected that evening.[11] Solomon and her staff enjoyed hosting these functions because they featured a set menu, decided upon by Solomon and the host. This simplified the order-taking process and made food preparation efficient and straightforward. Private gatherings occurred year-round, but they were most common during the December holiday season.

## THE COMPETITION

There were over 200 Italian restaurants among the thousands of restaurants in Toronto. These restaurants ranged from small, family-owned-and-operated establishments to national and international chains such as East Side Mario's and Alice Fazooli's, both part of a portfolio of restaurants owned by large investment groups. These establishments had multiple locations and ran national advertising campaigns. In addition to brand name, restaurants were classified by the type of food they served, with "pizza restaurants" being the most common.

---

[10] *Three bar/serving staff and two kitchen employees were paid for 12 hours of work on weekdays and seven hours on Saturdays.*

[11] *Tutti Matti's weekly sales breakdown was: Monday and Tuesday, 20 per cent of sales; Wednesday and Thursday, 30 per cent; and Friday and Saturday each contributed 25 per cent of weekly sales.*

With so much competition in Toronto, the sheer volume of dining options intimidated Solomon. Her main concern was that many restaurant-goers appreciated a familiar dining experience and, once they had an established group of preferred restaurants, they rarely tried new ones, unless encouraged — sometimes through advertisement, but most often by strong word-of-mouth. So far, Solomon had observed that over 90 per cent of her customers were returning diners, so she knew she was doing something right. She concluded that her major challenge would be to attract new customers whom she could convert to repeat diners. That said, there were two restaurants that Solomon identified as immediate competitive threats: Alice Fazooli's and Terroni. Both competitors had multiple locations and established reputations as popular Italian restaurants.

**Alice Fazooli's**

Alice Fazooli's (Fazooli's) opened in 1991 and was owned by SIR Corporation, also the owner of the Jack Astor's restaurant chain. Fazooli's offered Italian and Mediterranean dishes, most with considerable Canadian influence, and its menu offerings were comparably priced to Tutti Matti's. Fazooli's had grown to five locations in the Greater Toronto Area[12] and the chain's flagship location was a five-minute walk east of Tutti Matti, on Adelaide Street. Considerably larger than Tutti Matti, this restaurant had more than three times Tutti Matti's seating capacity. Fazooli's had recently launched a website where customers could view their menu and obtain contact information for each location. Located a block away from two of Toronto's leading theatres, Fazooli's on Adelaide was popular with the tourist and pre-theatre crowds. The four other locations were located in new commercial developments, near big-box stores, so they were popular suburban destinations. Recently, Fazooli's had been named a finalist in Toronto.com's *2002 Best of T.O.* annual Web survey.

**Terroni**

Terroni had been around for a number of years, with its third Toronto location having opened in 2000. Terroni specialized in thin-crust pizzas and was a popular lunchtime destination with its sandwich menu. Each location catered to a variety of customer groups. One of the Terroni restaurants was within walking distance from Tutti Matti, just outside of the financial district, which made it a convenient location for business lunches. In the evening, this restaurant was very popular with pre-theatre diners. Terroni had a website featuring its menu and contact information, and it enjoyed strong word-of-mouth advertising. Aside from its website, Terroni advertised minimally.

---

[12] The Greater Toronto Area (GTA) includes the city of Toronto, as well as four surrounding municipalities: Durham, Halton, Peel and York.

**Solomon's Goals for Tutti Matti**

Solomon was pleased with Tutti Matti's early performance, and she was on her way to completing a profitable first year of operations — a feat that was rare in the restaurant industry. After dealing with a number of staffing difficulties, menu adjustments, and having to pay for a few unexpected items, Solomon was eager to settle into a more predictable routine. Given the hearty nature of her menu, Solomon expected sales to be seasonal, with increased business during the colder winter months. Whether or not her predictions about seasonality were accurate, she was interested in options to encourage summertime sales. Also, as with most restaurants, the beginning of the week was slow, and Monday and Tuesday evenings were exceptionally slow at Tutti Matti. Since the majority of costs were fixed to keep the restaurant open, any increase in customers on these nights would have an immediate positive effect on the restaurant's profitability.

## FUTURE OPTIONS

### Advertising

Like many Toronto restaurateurs, Solomon found it very difficult to decide how to efficiently and effectively spend her advertising budget. There were many media choices for advertising, and the chosen methods would have to reach Tutti's target market and effectively convey the message. With so many start-up expenses,[13] the advertising budget for the first six months had been only $500. Solomon understood that establishing a strong customer base early was important, and, if justified, she was comfortable increasing her annual marketing budget up to four times for the upcoming year.

### Web Site

Over the past couple of years, an increasing number of restaurants had launched their own websites, and Solomon knew that people were increasingly using the Internet to investigate new dining options. While she was unsure what role a website would play in Tutti Matti's marketing strategy, Solomon wondered whether launching one would be a wise decision. She had been in touch with a local Web designer who would build a professional-looking site to accommodate her needs. Solomon would be interested in monitoring website traffic following specific promotions and in investigating the effect of media attention, such as editorial reviews, on the public's interest in Tutti Matti.

---

[13] *Included in Tutti Matti's expenses were miscellaneous start-up costs of $5,000 and utility costs, which typically worked out to five per cent of sales.*

Solomon was unsure whether spending on an online presence would be a worthwhile investment. Also, she was somewhat concerned as to whether a unique restaurant such as Tutti Matti, should, in principle, have a website, fearing that it might eliminate the best part of going to a new restaurant — the element of surprise. Solomon likened the ability to learn everything about a restaurant before ever setting foot inside it to knowing the ending of a great movie before entering the theatre.

If a site was launched, Solomon wondered what it should include. The Web designer had quoted a price of $1,500 for the site's creation, without e-commerce capabilities. Regardless of its content, the site would be hosted on a server provided by the Web designer at a monthly cost of $25. Solomon would employ one of her current staff, who had computer programming experience, to maintain the site, and she estimated two hours a month would be required to do so.[14]

**Summerlicious**

Solomon had just received an invitation to join the City of Toronto's Summerlicious Festival (Summerlicious). She was expected to respond within the next week. For two weeks each July, the Municipality of Toronto hosted Summerlicious, an opportunity to encourage visitors and residents of Toronto to dine out at the festival's featured restaurants. Historically, many of the city's most popular restaurants participated. The festival was held in conjunction with the Celebrate Toronto street festival, a weekend of free performances and activities held along Toronto's famous Yonge Street. This festival attracted thousands of people, both local and tourists, and each year it had enjoyed increasing popularity.

Only selected restaurants were invited to participate in Summerlicious. The city of Toronto funded a marketing campaign that featured public transit ads, billboards and ads in local newspapers and magazines in support of the festival. No particular restaurant was mentioned in the Summerlicious ads, but brochures listing all participating restaurants were available at multiple venues, including hotels, restaurants and information booths across the city.

To participate, Tutti Matti would have to pay a $350 registration fee. The Summerlicious organizing committee then classified each restaurant into one of two pricing levels – the first tier offered lunch for $20 and dinner for $35, and the second price offered lunch for $15 and dinner for $20. All restaurants had to offer a three-course meal, including an appetizer, a main course and a dessert, at the predetermined (*prix-fixe*) price. All new participants were placed in the second tier. If Tutti Matti participated, Solomon would have to offer menu selections from her regular menu because it would to too difficult to create new offerings for

---

[14] *Tutti Matti paid all serving staff a wage of $8 per hour. Kitchen staff received $12 per hour, and Solomon had negotiated a rate of $12 per hour for website maintenance.*

a two-week period. Solomon suspected that a Summerlicious customer differed from her typical customer. She had heard from other participating restaurateurs that alcohol/wine sales were cut in half during Summerlicious. This meant that the average bill per customer would likely decrease considerably. Solomon wondered whether there might be other concerns with Tutti's participation, or with being placed in the second tier. Summerlicious seemed like an interesting opportunity, and Solomon wondered how she should respond to her invitation.

**Bring Your Own Wine (BYOW)**

It appeared that the Municipality of Toronto could soon pass legislation allowing diners to bring their own bottled wine into Toronto restaurants and to take any unfinished home from the restaurant. Customers who brought their own wine could be charged a corkage fee, which would be determined independently by each restaurant. This practice was common in many European countries, and it had been in place in restaurants in Montreal, Quebec, for a number of years.

BYOW intrigued many restaurant owners, since it allowed them to collect corkage fees while incurring no incremental costs, and Solomon wanted to investigate the program's benefits. Corkage fees would range from zero to $35, regardless of bottle size or type, could vary by restaurant and, sometimes, even by night of the week. Solomon wondered whether to participate and, if so, what corkage fee would be best. She also questioned whether there was any merit to setting different corkage fees for different days throughout the week.

**Dining Out Magazine**

Solomon had recently been in contact with one of the editors of *Dining Out* magazine. *Dining Out* focused exclusively on restaurant editorials for various cities across North America. Each featured city had its own publication, and the Toronto edition was published three times annually and sold for $7 per issue. Solomon could purchase a one-page spread in the upcoming issue for $2,000. The suggested page layout would feature a brief history of the restaurant, a photo-shoot of selected menu items and the restaurant's interior and a list of selected menu items at the bottom of the page. Thirty thousand copies of each issue were printed and sold at retail outlets such as Chapters/Indigo. Copies were also given away at premium grocery outlets, such as Bruno's Fine Foods and Pusateri's, in the Greater Toronto Area. The editor told Solomon that each copy was viewed by an average of two people and, while she liked the idea of a restaurant-focused magazine, Solomon was somewhat concerned about her feature being overlooked with so many similar editorials in each issue.

**Toronto Life Magazine**

Another option under consideration was advertising in an upcoming issue of *Toronto Life* magazine. A monthly publication, *Toronto Life's* tagline was "Your Entertainment Source," and it was purchased by individuals and businesses alike. The magazine featured articles and reviews on everything from fashion to dining in Toronto. Each year, *Toronto Life* published a "Where to Eat" issue featuring reviews on the city's top and "up-and-coming" restaurants and restaurateurs. Solomon knew that the magazine had a healthy monthly readership, and she wondered whether an upcoming monthly issue or the next food issue would be a wise place to advertise. The Where to Eat issue came out in March, and a business-card-sized advertisement would cost $1,000.

**DECISION TIME**

Solomon knew that the next few months would be difficult due to decreased tourism and consumer spending, but she also knew that dining out remained an important part of many people's lives. Once the effects of the SARS outbreak faded, Solomon was confident that there would be a large upswing in restaurant spending, and she wanted to capitalize on this opportunity when it occurred.

With the interest on her loans coming due, Solomon was eager to continue the restaurant's financial success.

Tutti Matti needed a strong, creative marketing plan, soon. The restaurant had become Solomon's life, and she was willing to invest whatever time and effort was needed to build a solid foundation. While family and friends were still regulars at the restaurant, their appearances were becoming less frequent, and Solomon knew a more consistent customer base would be necessary to ensure success. Although she had spent considerable time researching her available advertising and promotional alternatives, perhaps there were other viable options that she had not considered.

## Exhibit 1

## MAP OF TUSCANY AND ITALY

Source: Italy Weather-and-Maps.com, http://www.italy-weather-and-maps.com/italy/maps/tuscanymap.php, accessed July 18, 2006. © italy-weather-and-maps.com 2003-2005, reproduced with permission.

## Exhibit 2

## MAP OF TUTTI MATTI'S LOCATION

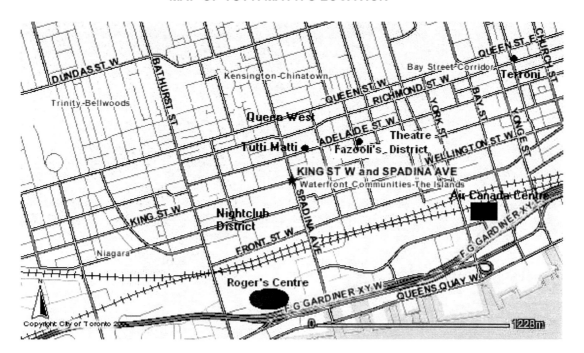

*Source:* Official Web Site of the City of Toronto, http://map.toronto.ca/imapit/iMapIt.jsp?app=TOMaps, accessed July 18, 2006. © City of Toronto, reproduced with permission.

## Exhibit 3

## PHOTOS OF TUTTI MATTI

**The Open Kitchen**

**The Front Dining Area**

**Exhibit 3 (continued)**

**The Front of Tutti Matti (From Adelaide Street)**

*Source: Ian DaSilva, July 16, 2006.*

## Exhibit 4

## TUTTI MATTI DINNER MENU

### ANTIPASTI

**Crostoni misti** Toasted country bread three ways: crostone with taleggio and quince jelly; crostone with three cheeses, speck, and mushroom pate; crostone with roasted artichoke and garlic   13.95

**Tris di prosciutto** Prosciutto di parma prepared three ways with country toasted bread, seasonal fruit and cheese   14.95

**Antipastio della casa** Insalata caprese with bufala mozzarella and chef's selection of imported cheeses, cold cuts, marinated vegetables and pâtés served with crostini   14.95

**Fagottini** Chickpea flour crepe stuffed with mixed mushrooms, asparagus and chef's mix of cheeses and truffle pâté drizzled with truffle honey   11.95

**Carpaccio di nana** Smoked duck carpaccio served with fresh orange, shaved pecorino di pienza and Tuscan extra virgin olive oil   12.95

**Carpaccio di manzo** Beef carpaccio served with fresh arugula, parmigiano, lemon and Tuscan extra virgin olive oil   9.95

**Carpaccio di trota** Cured rainbow trout carpaccio served with fresh red onion, fennel and red beet vinaigrette   11.95

### INSALATE

**Insalata dei Matti** Fresh arugula with grapes, walnuts and shaved pecorino di pienza with pomegranate vinaigrette   9.95

**Insalata della principessa** Baby spinach with bufela mozzarella, pine nuts, sliced prosciutto and dried figs served with fruit caramel vinaigrette   10.95

**Insalata verde** Mixed greens served in a balsamic vinaigrette   6.95

**Insalata di barbabietola** Variety of roasted beets on a bed of fresh arugula served with horseradish vinaigrette   12.95

### PRIMI

**Pasta del Giorno** Pasta of the Day   priced daily

**Tagliatelle con funghi** Fresh saffron tagliatelle with wild mushrooms, truffle pâté, garlic and Tuscan extra virgin olive oil   17.95

**Pinci alla boscaiola** Long hand-rolled pasta served with a stewed wild mushroom and artisanal sausage   18.50

**Papardelle con stracotto** Hand-cut pappardelle with pulled brisket, cherry tomatoes, garlic and fresh herbs   16.00

**Chitarrini con rucola e capra** Fresh guitar string pasta with arugula pesto, chevre and marscapone   16.95

## Exhibit 4 (continued)

### SECONDI

*Served with seasonal vegetables*

**Pesce del giorno** Fish of the day    priced daily

**Tagliata** 12oz. Black Angus strip loin served with fresh arugula, lemon, extra virgin olive oil and baby onion marmalade    28.95

**Costole di maiale-- "Nastro Azzuro"** Pork short ribs slow roasted with Nastro Azzuro Italian beer, blood oranges, rosemary and garlic    20.95

**Vitello al tartufo salvi** Pan seared veal tenderloin with truffle pâté and sage    23.95

**Scottiglia cinghiale** Wild boar stewed with red wine, juniper berries, bay leaves and rosemary served with fagioli ucceletta (cannellini beans stewed with tomatoes, garlic and sage.)    18.95

**Arrosto del giorno** Game/meat of the day    priced daily

### CONTORNI

**Spinaci all'aglio** Sauteed spinach with garlic    4.00

**Cipolline in agro dolce** Sweet pickled baby onions in balsamic reduction    4.00

*Source: Tutti Matti, July 18, 2006.*